Alberto P. Guimarães

From Lodestone to Supermagnets

Understanding Magnetic Phenomena

Alberto P. Guimarães

From Lodestone to Supermagnets

Understanding Magnetic Phenomena

WILEY-VCH Verlag GmbH & Co. KGaA

Alberto P. Guimarães
Centro Brasileiro de Pesquisas Físicas
Rio de Janeiro, Brazil

Library of Congress Card No.: applied for

British Library Cataloging-in-Publication Data: A catalogue record for this book is available from the British Library.

Bibliographic information published by Die Deutsche Bibliothek
Die Deutsche Bibliothek lists this publication in the Deutsche Nationalbibliografie; detailed bibliographic data is available in the Internet at http://dnb.ddb.de

Typesetting Typomedia GmbH, Ostfildern
Printing and Binding Ebner & Spiegel GmbH, Ulm
Cover Design Himmelfarb, Eppelheim, www.himmelfarb.de

Printed in the Federal Republic of Germany

Printed on acid-free paper

ISBN-13: 978-3-527-40557-2
ISBN-10: 3-527-40557-7

For Ricardo, Marília and Silvia.

This popular science book presents basic facts on magnetism. Alberto P. Guimarães, author of *Magnetism and Magnetic Resonance in Solids* (Wiley 1998) and a working scientist in the area of magnetism, is also active in the popularization of science, as one of the founders of the science magazine *Ciência Hoje*. He obtained his PhD at the University of Manchester, England, and is now professor of physics at the Brazilian Center for Research in Physics (CBPF) in Rio de Janeiro.

He describes these topics in a way that is easy to understand. He starts before the Greeks and the first records of magnetism and ends up with modern supermagnets. Guimarães emphasizes the evolution of the great ideas of physics, relating them to the background of the general evolution of science as a whole. He also focuses on the evolution of magnetic materials and how they are changing our lives.

This description of the discovery and history of magnetism enables a person of non-scientific background to understand the historical development of some of the important scientific ideas of magnetism, a topic that plays a significant role in modern life, from hard disks to magnetic resonance imaging.

Table of Contents

From Lodestone to Supermagnets. Alberto P. Guimarães

ISBN: 3-527-40557-7

"A wonder of such nature I experienced as a child of 4 or 5 years, when my father showed me a compass. That this needle behaved in such a determined way did not at all fit into the nature of events, which could find a place in the unconscious world of concepts (effect connected with direct "touch"). I can still remember – or at least believe I can remember – that this experience made a deep and lasting impression upon me. Something deeply hidden had to be behind things."
Albert Einstein, Autobiographical Notes*.

* A. Einstein, Autobiographical Notes, in *Albert Einstein: Philosopher-Scientist*, vol. I, Ed. P. A. Schilpp, Cambridge University Press, London, 1969, p. 9.

From Lodestone to Supermagnets. Alberto P. Guimarães

ISBN: 3-527-40557-7

Preface

The fascination of magnets dates back some three thousand years, or possibly even more. The earliest accounts attribute to the ancient Greeks the first observations of magnetic phenomena. Studies of the Middle Ages and of the 16th and 17th centuries, inspired by the image of a magnet attracting pieces of iron, have stimulated many insights into the workings of nature, including the model for the forces that hold planets in their orbits.

The quest for the understanding of the magnetic forces is a remarkable story in itself. This comes about in the first place through the discovery of the connection between electricity and magnetism in the 19th century, a breakthrough that also paves the way for the invention of every known piece of electrical machinery. A second decisive step is the development of Quantum Mechanics, which allows the understanding of the structure of matter, and of magnetic order. This basic understanding, combined with advances in the knowledge of materials, led to the production of better magnets, making them ubiquitous in modern life.

This story unfolds against a background of evolving ideas on nature and on the possibility, and the best strategy, for acquiring knowledge about it. It spans a period that has witnessed the birth of science, the scientific revolution of the 16th and 17th centuries, and the turning points in world view that have shaken the Newtonian system, and ultimately challenged long-established ideas such as causality and determinism.

The process of accumulation of scientific knowledge during the last 3000 years is presented here as an integral part of the fascinating progression embodied in the history of science.

From Lodestone to Supermagnets. Alberto P. Guimarães

ISBN: 3-527-40557-7

I am grateful for the comments and suggestions made by Bill D. Brewer, Ivan S. Oliveira, Ximenes A. Silva and Cássio Vieira, and to many other friends and colleagues who read and commented on parts or the whole of the manuscript: A. M. O. Almeida, W. Baltensperger, G. Bemski, T. J. Bonagamba, S. R. Bregman, M. A. Continentino, G. O. Corrêa, L. Davidovich, D. M. Esquivel, A. M. Gomes, N. Kaplan, M. Knobel, R. Lent, C. H. Lewenkopf, M. L. Maciel, H. Micklitz, R. P. A. Muniz, A. M. Oliveira, L. A. de Oliveira, P. Panissod, L. C. Sampaio, A. F. F. Teixeira, M. B. Bulcão and J. Wisnovsky. I am also grateful to those who obtained some elusive references: V. S. Amaral, W. D. Brewer, M. Z. P. Guimarães, J. Huehnergard, E. Ihrig, M. Knobel and R. S. Sarthour. I am indebted to Andressa Furtado and Márcio Paranhos for the drawings.

I would also like to thank the Bakken Museum, of Minneapolis, for a Visiting Research Fellowship to consult their rich collection and search for iconography on magnetism.

For further material to supplement this book see www.cbpf/lode stone.

Alberto P. Guimarães

Chapter 1

A Stone with a Soul

The secret of magnets, now explain that to me!
*There is no greater secret, except love and hate.**

Johann Wolfgang von Goethe (*Gott, Gemüt und Welt*)[1]

Magnetism – The First Records

What child has not played with a magnet, amazed at its power to attract iron objects? Who has not felt intrigued with its quasi-magical ability to exert forces at a distance? Forces that appear to act without material mediation, without physical contact; one does not have to touch a piece of iron to move it with a magnet, in contrast with all other usual physical experiences**.

The observation of magnets has accompanied humankind for more than three thousand years, for these wondrous objects were known before the first millenium BC. In ancient Mesopotamia, iron oxides were used in weights from the late third millenium BC[2]. Different iron ores, including the mineral magnetite, were used in the same region to make seals ('cylinder seals') from 2000 BC.

The earliest indication that magnetic phenomena were known in the ancient world is the fact that the magnetic mineral magnetite, the naturally occurring magnetic stone, was referred to in Mesopotamia as 'grasping hematite' or 'hematite that seizes' (shadânu sabitu)[3,4,5]. This term is used in a tablet with a list of commodities[6] from the first half of the second millennium BC, giving an indication that this remarkable property of some iron ores had been observed very early. The expression 'grasping hematite' is also found in the 16th tablet that is

* "Magnetes Geheimnis, erkläre mir das!
Kein grösser Geheimnis als Lieb und Hass".

** Except for the (usually weaker) attraction by rubbed amber, which we will consider in later chapters.

From Lodestone to Supermagnets. Alberto P. Guimarães

ISBN: 3-527-40557-7

part of a series called Har-ra hubullu, containing Sumerian words and the Akkadian equivalent, from the first millenium BC[7]. The phrase 'living hematite' is also recorded.

The first known records of the properties of the magnet were made in Greece. The Greek philosopher Thales of Miletus, who lived in the 6th century BC, was said to have considered that the magnet had a soul.

The origin of the name magnet, according to the Roman poet-philosopher Lucretius (Titus Lucretius Carus) (c. 98–55 BC) writing in the 1st century BC in the didactic poem *De Rerum Natura* ('On the Nature of Things'), was in the province of Magnesia, in Thessaly, northern Greece. Lucretius wrote[8]: "Next in order I will proceed to discuss by what law of nature it comes to pass that iron can be attracted by that stone which the Greek call the Magnet from the name of its native place, because it has its origin within the bounds of the country of the Magnesians." Inhabitants of Thessaly colonized Asia Minor, where there are two provinces with the name Magnesia. Since the lodestone is referred to as 'Lydian stone', in some accounts, including one[9] of the Greek poet Sophocles (496–406 BC), one may relate the discovery to the Magnesia in the region of Lydia[10].

A different account is given by Pliny the Elder (Gaius Plinius Secundus) (AD 23–79), a Roman author who wrote the encyclopedic 'Natural History' (*Historia Naturalis*), a thirty-seven-volume treatise whose influence lasted for some 15 centuries. Pliny compiled many thousands of observations on natural phenomena, plants, animals and places, from his own experience and from more than 2000 earlier texts. He was untiring in this task, studying and writing throughout his whole life. His curiosity finally led to his death as he approached the volcano Vesuvius to observe more closely the eruption of the year AD 79 that destroyed Pompeii. He mentions the property of magnets of attracting iron, and informs that according to Nicander (a Greek poet), "it was known as *magnes* after its discoverer, Magnes, who found the mineral on Mount Ida*. (...) The discovery is said to have been made when the nails of Magnes' sandals and the tip of his staff stuck to the stone as he was grazing his herds."[11]

It is difficult to establish which of the traditional versions on the

* There is a Mount Ida in Asia Minor (Turkey); in the Illiad, Zeus watched the Trojan War from Mount Ida, also known locally as Kaz Dagi.

origin of the word 'magnet' presented above is more reliable. In any case, we know that the first objects to show this amazing property were rocks containing iron oxides, a mineral known as magnetite, basically an iron oxide of formula Fe_3O_4, brown or black, with a metallic luster. The magnetite that behaves as a magnet is known as lodestone, from the word lode, archaic English meaning course or way.

The knowledge of magnetism and magnetic materials was acquired very slowly, from the beginnings of science. Knowledge about minerals had been accumulated since prehistoric times, as Man observed and classified natural materials according to their physical properties, used them, and at a later stage tried to tailor them to his practical needs. The pace of accumulation of knowledge accelerated as the first human groups started to settle, abandoning nomadic life, and began to raise the first crops in Mesopotamia, the region between the rivers Tigris and Euphrates in present-day Iraq, and also along the river Nile, in Egypt. These civilizations learned how to grow wheat and barley, raise cattle, work metals like copper and bronze (an alloy of copper and tin). Iron was initially extracted from metallic meteorites; iron extracted from ore appears in the first half of the third millenium BC[12]. Their way of life required a certain amount of botanical knowledge, knowledge of the cycle of seasons, and physical knowledge about the melting points of the metals, their degree of hardness, malleability, and so on.

The development of the reckoning of time was related to the growth of astronomical knowledge. Time was naturally measured in days, the period from sunrise to sunrise (corresponding to the rotation of the Earth around its axis); a longer period was recognized as the year, the interval from one cycle of seasons to the next (corresponding to the time of revolution of the Earth around the Sun). The determination of how many days were contained in a year led in Mesopotamia and Egypt to the introduction of a calendar that incorporated the yearly periodicity of the seasons, important for the agricultural society; it was convenient to start the seasons on the same day, every year. This demanded the observation of stars, and in general, the acquisition of astronomical data. The division of the day into hours was not so relevant for life in a primitive society, the activities being limited from dawn to dusk. Also, measurement of time on this shorter scale involves creating new instruments and solving many technical problems.

The construction of housing and temples, and the need to measure land led to the development of units of measurement; a system of weights and measures was created in Mesopotamia as early as 2500 BC, and later also in Egypt. The usual units of length were related to parts of the human body: the cubit, the distance from a man's elbow to the fingertip; the span, the length of a fully stretched hand from the tip of the thumb to the tip of the little finger, and so on. One of the earliest units of weight used in Mesopotamia was the mina; another unit was the shekel, equivalent to 1/60 of the mina. The shekel was equivalent to the weight of 120 or 200 grains of wheat[13].

A giant step in the development of early human society was the invention of writing by the Sumerians, one of the peoples of Mesopotamia. This started as a pictographic representation, where an ox was represented by a picture of the head of the animal, and later took a more abstract form, in the so-called cuneiform language, around 3000 BC. This name is given because of the shape of the short strokes used (from the Latin *cuneus*, 'wedge'). The next stage in the evolution of writing was the shift from representation of things to representation of sounds.

Thales and the Beginnings of Greek Science

The beginnings of science are usually placed in Greece, in the 6th century BC. There is a certain degree of arbitrariness in this choice, since scientific and technical developments were, of course, also made in Egypt, Babylonia and China. However, it is generally accepted that the kind of knowledge, or rather the approach towards knowledge itself, of the first investigators in Greece had the seeds of the scientific endeavor*.

Why Greece? Why did it not occur in Egypt? It is known that in Egypt, despite the wealth of practical knowledge accumulated, there is no evidence of systematic theorizing about natural phenomena. In Egypt, the identification of the gods to the kings, with the priests being

* "To the extent that it is explicative and ontological [i.e. related to the being], science is a creation of the Greek genius – and if one considers these two aspects as essential, one can assert that science was born in Greece." (R. Taton, *Histoire Générale des Sciences*', vol. I, Presses Universitaires de France, Paris, 1966, p. 204).

both religious and secular authorities, did not contribute to creating a climate that stimulated free speculation about the universe[14]. In China, on the other hand, law did not guarantee individual rights; customs and a 'natural law' prevailed, without the existence of prescribed sanctions for different crimes[15]. It has been speculated that since there was no immediate connection in China between human acts and their legal sanctions, the relationship between the natural phenomena and general laws that governed them was not easily perceived; the concept of laws of nature, therefore, was not valued[16].

In the Greek world, one of the elements that contributed to the emergence of science was the fact that the gods were not associated with secular power, and as a consequence, they could be more easily removed from the privileged role of major actors that determined the course of natural events. In the early mythical accounts, nature tended to be endowed with the power to evolve from within, a useful characteristic to open the way for the causal explanation of natural phenomena[17]. The creation of the universe, as described in Hesiod's Theogony, is not the act of gods, but rather resulted from the action of "relatively abstract entities", such as Chaos, Earth and Eros[18] (Theogony means an account of the origin and descent of the gods).

Political factors have also been pointed out as relevant for this development: the Greek city-states enjoyed a variety of constitutional forms of organization, among these the novel democratic regime, which allowed the development of an atmosphere of free speculation. Indeed, in the case of Athens, the constitution permitted a high level of participation in the political life of the city[19]. The extent of the maritime trade, with the exchanges between different cultures, particularly with the Egyptians and the peoples of Mesopotamia, may have worked in favor of expanding the horizons of the first thinkers. A higher level of literacy in Greece, as compared with Egypt and Babylonia, was another favorable factor[20].

The Greek world in Antiquity was constituted of cities scattered on the edges of the Mediterranean Sea (Figure 1.1); Egyptian civilization was concentrated on the banks of the Nile, relatively isolated from other societies by the desert, east and west of the river. Both in Egypt and in Greece, however, intellectual life could develop only when society spared some of its members from productive activities, allowing them to enjoy free time for leisure, as first pointed out by the philosopher Aristotle: "Hence, when all such inventions were already estab-

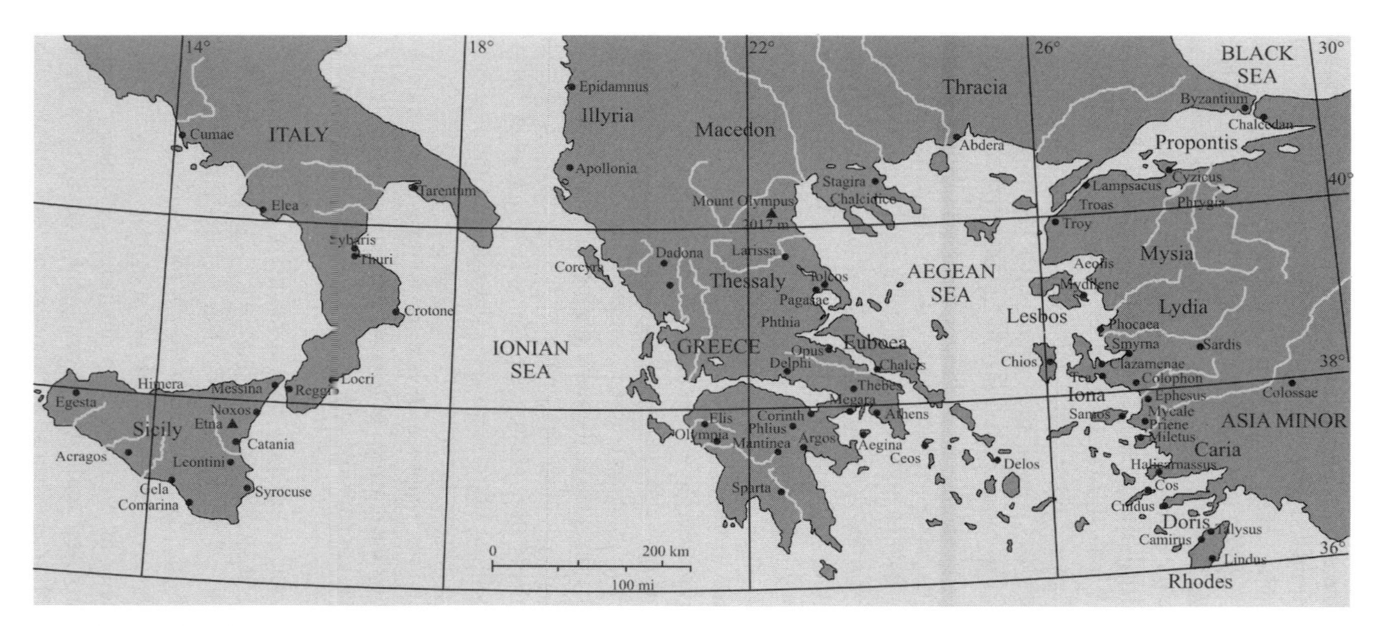

Figure 1.1 The Greek world.

lished, the sciences which do not aim at giving pleasure or at the necessities of life were discovered, and first in places where men first began to have leisure. This is why the mathematical arts were founded in Egypt; for there the priestly caste was allowed to be at leisure."[21] And furthermore "(...) for it was when almost all the necessities of life and the things that make for comfort and recreation had been secured, that such knowledge began to be sought."

Around 1400 BC the first writing system based on the sounds of the words was invented in Greece. This evolved some centuries later to become an alphabetic system, where syllables were formed with consonants and vowels[22].

In the 6th century BC there appeared in written form the Iliad and the Odyssey, epic poems attributed to Homer, a poet-singer who would have lived some centuries before, possibly on Chios, an island in the coast of Asia Minor or Anatolia, the Asian portion of Turkey. Initially transmitted through oral tradition, these masterpieces of Greek literature were centered on the adventures of the legendary characters Achilles and Ulysses (Odysseus); in the Iliad, the history of the Trojan War is told (Ilion is one of the names of Troy). Besides their great literary value, these works are important as they reveal a certain knowledge of natural phenomena, of medical and agricultural practices prevalent in Greece at that time[23]. The Iliad and the Odyssey helped to consolidate the Greek language, and became part of the education of every literate Greek for many centuries. Two other poems, the *Theogony*, and *Works and Days*, are attributed to Hesiod, who may have lived in Ascra, in Boeotia, mainland Greece, around 700 BC. Herodotus, the Greek historian who lived in the 5th century BC, expressed the importance of Hesiod and Homer in shaping Greek culture with the following words: "It is they who created a theogony for the Greeks, gave the gods their names, distributed their privileges and skills, and described their appearance."[24] Or, in the words of Xenophanes, a philosopher and religious thinker from Colophon, in Ionia (the central part of the coast of Asiatic Turkey), who lived in the sixth to 5th century BC: "What all men learn is shaped by Homer from the beginning."[25] The poems of Hesiod and Homer, that mingled myth with practical knowledge, mark the transition of Greece into a new intellectual era.

This era opened in the 6th century BC, as some inquisitive men started to ask new questions about the world that surrounded them, and furthermore, tried to search for answers that did not contain the usual elements of lore and mythology; this meant trying to find natural causes for the natural phenomena. In these first bold endeavors lie the origins of scientific thought. These men were the philosophers, or lovers of wisdom, in Greek; Aristotle would later on refer to them as *physikoi*, from *physis*, nature. From their standpoint, they avoided the attitude common in pre-scientific societies, of mistaking "association of ideas in their minds" for "causal relations between things", in the words of the nineteenth-century British anthropologist James Frazer[26]. First of all, they began to view the world not as some disordered, or arbitrary, aggregate of things or sequence of events, but as an object of knowledge and explanation. The word *kosmos*, from a verb meaning 'to order', was used to describe the universe; the very fact that a word was required to designate the totality of things represents in itself an important move in this new direction. Their attitude is summed up in the words of W. K. C. Guthrie[27], author of *A History of Greek Philosophy*: "Philosophy started in the faith that beneath this apparent chaos there exists a hidden permanence and unity, discernible, if not by sense, then by the mind."

It is very difficult for us nowadays to apprehend the magnitude of the change in world view that the attitude of the early philosophers entails, since so much of their contribution has become a tacit part of our intellectual inheritance*. We are used to searching for natural causes of phenomena without having to make the conscious choice between a mythical and a scientific standpoint; under most circumstances, the choice between *mythos* and *logos* is immediate.

The first philosophical ideas appeared in the eastern and in the western frontiers of the Greek world (Figure 1.1): in the East, along the cities of the coast and in the islands of Asia Minor, and in the West, in Sicily and mainland Italy. Very little original written work or even fragments of works of the early philosophers has survived to our days. Most of their ideas reach us through the writings of commentators or historians, who lived centuries after them, and who did not have access to the original manuscripts; to make things worse, even these

* The philosopher Karl Popper places the originality of the first philosophers not so much in the fact that they abandon the myths, but in their critical attitude: they substitute the "traditional preservation of the dogma", for a "tradition of criticizing theories" (K. R. Popper, *Objective Knowledge*, Oxford University Press, Oxford, 1975, pg. 348).

later accounts were preserved in the form of copies of copies of texts[28].

The first of these thinkers is Thales of Miletus (c. 640–546 BC), who is regarded as the founder of Greek science and philosophy. He made the daring move of proposing that there was something common to the entire universe; in this first unified view of the universe, he posited, according to Aristotle*, that "all things are water". The choice of water as the element that formed everything in the universe may have arisen from the fact that water can present itself as a solid, as a liquid, and as a gas. This choice for the prime element of the universe may also have resulted naturally from the importance of water in the life of animals and plants; the relevance of the sea for the economic activity of the peoples in the region may have also played some role. This idea had antecedents, since many early myths involved sea and water deities, and Babylonian cosmology around perhaps 2000 BC already assumed a primacy of water[29]. It was also part of these myths that the Earth itself was regarded as floating on an infinite pool of water.

Thales thus proposed the first vision of the universe that lacked mythological elements, gods or demons, as essential constituents. He had a reputation as a mathematician, and he was the first man to be considered responsible for specific mathematical discoveries[30]. However, some of his alleged scientific exploits are not generally accepted today, such as the famous prediction of a solar eclipse[31] in 585 BC; it is argued that there simply was not sufficient astronomical knowledge at the time to allow such a prediction.

Two other thinkers constitute, with Thales, the School of Miletus: they are Anaximander (610–550 BC) and Anaximenes (550–486 BC). Anaximander elaborates on the world view of Thales, assuming a single element as the principle of everything, but in his view, this was an abstract element, the *apeiron*, or infinite. This choice represents a more sophisticated attempt at the comprehension of the principle, or *physis*, of the universe, since it shifts from known material substances to a pure abstraction. Simplicius, a commentator and philosopher of

* "Most of the first philosophers thought that principles in the form of matter were the only principles of all things. (...) Thales, the founder of this kind of philosophy, says that it is water (that is why he declares that the earth rests on water)." Aristotle, Metaphysics 983b6–11, 17–27, in J. Barnes, *Early Greek Philosophy*, Penguin, London, 1987, p. 63.

the Neoplatonist school, who was born in the second half of the 5th century AD, wrote[32] of Anaximander: "Of those who hold that the first principle is one, moving and infinite, Anaximander, son of Praxiades, a Milesian, who was a successor and pupil of Thales, said that the infinite is principle and element of the things that exist. He was the first to introduce this word 'principle'. He says that it is neither water nor any other of the so-called elements but some different infinite nature, from which all the heavens and the worlds in them come into being. And the things from which existing things come into being are also the things into which they are destroyed, in accordance with what must be."

The philosopher Anaximenes, in his turn, chooses air (or breath) as this prime element. He proposes mechanisms of condensation and rarefaction as the processes of formation of the variety of qualities and material aspects of the universe. Condensation of air produces cold, further condensation creates wind, cloud, rain, earth and rock. In an ancient text falsely attributed to Plutarch[33], one reads: "Anaximenes, son of Eurystratus, a Milesian, asserted that air is the first principle of the things that exist; for everything comes into being from air and is resolved again into it. For example, *our souls,* he says, *being air, hold us together, and breath and air contain the whole world* ('air' and 'breath' are used synonymously)".

Another great philosopher of importance in the history of science was Pythagoras; he was born on Samos, in Ionia, on the island closest to the continent, probably about the year 570 BC. To flee from the tyranny of Polycrates he migrated to Croton, in Southern Italy, where he established his school. He was then forty. His presence in Croton coincided with a period of increase in the influence of this city-state. At the end of the 6th century, after a rebellion in Croton, Pythagoras migrated again, ending his days in Metapontum, also in Italy, in the Gulf of Tarentum.

Besides being a philosopher and a political leader, Pythagoras created a religious sect; the facts of his life are entangled with many myths concerning his person. There is no certainty that Pythagoras left written works; a veil of secrecy covered his religious society, where members were obliged to take a five-year vow of silence[34]. His philosophical work was apparently accessory to his religious interests[35]: "What we may safely say is that for Pythagoras religious and moral motives were dominant, so that his philosophical inquiries were destined from the start to support a particular conception of the best life and fulfil certain spiritual aspirations."

For the Pythagoreans, numbers had an independent existence, and were more important than the natural phenomena; these were important insofar as they reflected the pre-eminence of numbers[36]. This supreme position of the numbers was described by the philosopher Aristoxenus (fl. 4th century BC) as having arisen from Pythagoras' interest in the practical side of commerce; Aristoxenus also considered him responsible for the creation of a system of weights and measures.

One of the first authors to write on the Pythagoreans was Aristotle, born in 384 BC; commenting on their relation with numbers he affirms[37]: "since, then, all other things appeared to have been modeled on numbers in their nature, while numbers seemed to be the first things in the whole of nature, they supposed that the elements of numbers were the elements of all the things that exist, and that the whole heaven was harmony and number". Although Pythagoras and the Pythagoreans are nowadays believed to have contributed little to the techniques of mathematics[38], the importance of their contribution to the conceptual foundations of mathematics is undisputed. One of these contributions is the fuller realization of the abstract nature of numbers and geometric figures, and the recognition that theses are entities that belong to a class separate from that of physical objects.

Before the appearance of the first philosophers, Egyptians and Babylonians had a practical knowledge of mathematics, which was used in trade and in keeping the accounts of the temple, and for the measurement of agricultural land. Although arithmetical operations with numbers were used, the meaning of the numbers was not discussed or examined in depth. The development of the number concept is related to the Pythagoreans; although numbers were used in reckoning from much earlier ages, the use of expressions such as 'two oxen' did not imply knowledge of the full meaning of the word 'two'. In other words, mathematical concepts such as the number concept could only be understood when one abstracted something that was common in 'two oxen' or 'two people', and so on. The same process involves the geometrical concepts: a rectangular table and a rectangular plot of land have something in common that has to be identified as the rectangular shape; in this way, the concepts of rectangle, square, circle, etc. were born. Geometrical properties of simple figures were known, including the fact that in a right triangle the sum of the squares of the lengths of the sides adjacent to the right angle is equal to the square of the length of the opposite side, a theorem known nowadays as 'Pythagoras' theorem'.

The development of the number concept and the importance of numbers in the Pythagorean world view have been associated with the introduction of coins, which first appeared in Lydia, in Anatolia, in the 7th century BC. One consequence of the emergence of a monetary economy, with the substitution of concrete goods by their abstract representation as numerical values in monetary units, may have been a stimulus for the Pythagorean discovery of the importance of numbers. The universalization of a system of weights and measures contributes in the same direction, helping to reveal the number as an abstraction. Measuring the weight of different goods with the same unit, for example, allows the same kind of association provided by currency: concrete goods of different nature may have the same weight, i.e. may correspond to the same number. In ancient Greece many of the units were the same as those used in Egypt and in the east; the cubit as unit of length, and the talent, unit of weight corresponding to about 58 pounds (25.8 kg) are examples. These units are already found in the verses of Homer, and were therefore in use by the 7th century BC.

Although the abstract nature of numbers was then recognized, numbers still retained a certain connection to physical objects; this appears, for example, in the representation of numbers by the Pythagoreans with pebbles or dots[39] (Figure 1.2). Among numbers of special interest were those that could be represented by a triangular array of dots, like 1, 3, 6, 10 and so on, called triangular numbers (the number 1, of course was a 'point-like' triangle). There were also the square numbers: 1, 4, 9 and so on. Some remains of this vision of the numbers have survived to our days, in the naming of the second power of a number as 'square' of the number, and the third power as 'cube'.

Pythagoras, or the Pythagoreans, established, probably using a monochord instrument, the *kanon*, that the musical intervals corresponded to some prescribed lengths of the string, which was varied by changing the position of a moveable bridge. The frequency of the musical notes was inversely proportional to these lengths. Thus, the intervals of Greek music corresponded to the frequency ratios 1:2 (octave), 3:2 (fifth) and 4:3 (fourth), again confirming to them that the

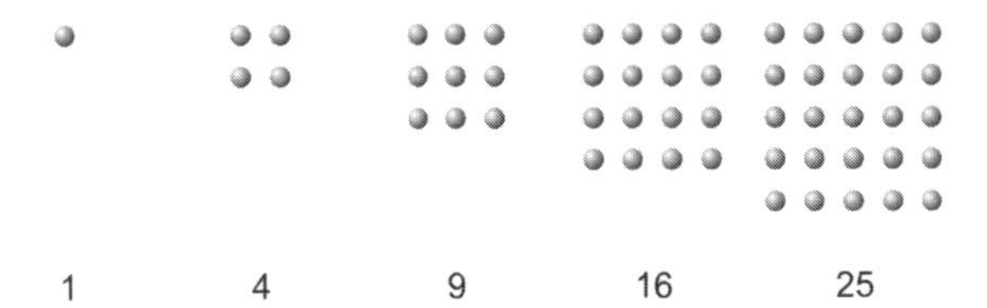

Figure 1.2 Pythagorean representation of square numbers: $1 = 1^2$, $4 = 2^2$, $9 = 3^2$, $16 = 4^2$, $25 = 5^2$.

first four integers 1, 2, 3 and 4 played a fundamental role in the world order[40].

Plato and Aristotle

One of the most important Greek philosophers was Plato, one of the greatest figures of the history of ideas. He was born around the year 428 BC, and came from an important family in Athens; the young Plato was an admirer of Socrates, a friend of the family and a major Greek thinker. Socrates had discussed the possibility of Man attaining knowledge, and also how to lead a life according to ethical principles. Socrates had shifted his mind from the themes of interest of the preceding philosophers, in cosmology and the physical world, to the problems that had Man at their center. One may assert schematically that he moved from the problem of the reality of nature, which was at the center of the investigation of the Milesians, to the problem of man; this steered him into an ethical inquiry that would ultimately take his life. Socrates was tried in 399 BC and condemned to die by poisoning with hemlock for his questioning of the Greek ruling authorities. This episode shook Plato and further stimulated him to follow his own path as a philosopher.

Plato's writings were in the form of dialogues, and have survived to this day as pieces of great philosophical and literary value. Plato established his school around the year 387 BC outside Athens; the institution became known as the Academy, apparently because it was established in a place named after the hero Hekademas.

Discussing the possibility of knowing the external world, Plato[41] "thought that the reason for our ability to know the outer world is that the same *simplicity* and *order* that please us in our *ideas* are also found in the *objective world.*" Plato thought that the objects of knowledge did exist, although they were not to be identified with anything in the perceptible world[42]. They were the forms, or ideas, of which the concrete objects of the world were but imperfect copies. In one of the dialogues, the *Meno*, Plato argues that knowledge is in fact a process of recollection; the immortal soul knows, for example, geometry, and this can be shown by properly questioning even an illiterate person.

Another giant among the Greek philosophers who left his mark in the birth of science was Aristotle. Aristotle was born in Stagira, presently Stavró in the north of Greece, on the Chaldice peninsula, in the year 384 BC. At the age of 18, he entered the school of Plato in Athens, where he remained until Plato's death. He then moved from Athens to Mysia and after a while to Mitylene, on the island of Lesbos. In 342 or 343 BC King Philip of Macedonia, who had been educated in Greece, invited him to Pella, capital of the kingdom, to tutor his son Alexander, then a 13-year-old boy. Alexander would grow up to become Alexander the Great, the king who would lead the Macedonians to conquer Persia, Babylonia, Egypt and extend their empire to India. Aristotle took this position, and after eight years, when Philip died, returned to Athens and founded his own school, where he lectured on rhetoric, sophistic and politics, and for a smaller number of students, on logic, physics and metaphysics. He used to discuss with the students while he walked, and from this fact, the school became known as the 'Peripatetics', (from the Greek *peripatein*, to walk up and down).

Aristotle left an extensive written work that can be divided into three parts: the scientific treatises, some separate texts related to those, and works of a more popular character. Among these treatises, the most important is the group of works known as the *Organon* ('Tool'), then the natural history works such as the *Historia Animalium*, as well as the Metaphysics, the Nicomachean Ethics, and the Politics. Aristotle chose for his works not the form of dialogues, but that of lecture notes. He classified theoretical knowledge into a) natural philosophy, b) mathematics, c) first philosophy (metaphysics).

In the first part of his treatise of logic, the Prior Analytics, and in the second, called the Posterior Analytics, he discusses the syllogism, with the goal of establishing the conditions for scientific knowledge[43]; with this work Aristotle set the foundations of formal logic. The syllogism, in its simplest form, is a series of three propositions, in which the first two are premises that lead to the third, the conclusion. The classic example is 1) "All men are mortal"; 2) "Socrates is a man"; 3) "Therefore Socrates is mortal". If the first two statements are valid, the third (conclusion) follows. This is a special case of deduction, or deductive reasoning, seen by Aristotle as a reasoning that goes from the general to the particular (a view not accepted nowadays by most philosophers[44]). Induction, on the other hand, goes from the particular to the general*; deduction and induction are two logical processes essential in the scientific discourse.

Aristotle valued observation and to a certain extent experimentation; he developed his doctrine starting from his knowledge of biology. On account of the interest in biological studies of the members of his school – the Lyceum – he kept a collection of zoological specimens, obtained in part from the travels of Alexander.

Aristotle's way of constructing his science is exemplified by Gillispie[45], speaking of his physics: "It was a serious physics, a consistent and highly elaborated ideation of natural phenomena. It started from experience apprehended by common sense, and moved through definition, classification, and deduction to logical demonstration. Its instrument was the syllogism rather than the experiment or the equation. Its goal was to achieve a rational explanation of the world by showing how the myriad subordinate means are adapted to the larger end of order."

He considered the existence of four different causes, using the word in a different sense than it has today; of these, the final cause, he called "prime mover". This prime mover was often called God, but it represented more a natural principle than a religious God; it was not a source of force, but an object of desire[46]. For example, the stars turn in the sky because of their desire to reach perfection, which is attained with their circular motion.

The point of view of Aristotle and followers contrasts with the vision of Plato. In the words[47] of the science historian Alexandre Koyré (1892–1964): "If you claim for mathematics a superior status, if more than that you attribute to it a real value and a commanding position in Physics, you are a Platonist. If on the contrary you see in mathematics

* Aristotle usually speaks of induction, however, not progressing from individuals (e.g. "this man", "this horse"), but from species ("man", "horse") to genus (i.e. "mammals", "living beings") (D. Ross, *Aristotle*, Methuen & Co, London, 1974, p. 39).

an abstract science, which is therefore of a lesser value than those – physics and metaphysics – which deal with real being; if in particular you pretend that physics needs no other basis than experience and must be built directly on perception, that mathematics has to content itself with the secondary and subsidiary rôle of a mere auxiliary, you are Aristotelian."

Another important philosophical movement that flourished in Athens was that of the Stoics; they were a group of philosophers that got this name from the place where they used to meet, the *Stoa poikile*, or Painted Porch. They were influenced by the ideas of the pre-Socratic philosopher Heraclitus (who was active around 500 BC) and their most important leader was Zeno of Citium (c. 334–262 BC), who was born on Cyprus and founded the school around 300 BC. They regarded the universe as a living being, filled with a fluid, the *pneuma*. They believed all matter to be formed from the four elements: earth, water, fire, and air. The Stoics were responsible for advances in logic and linguistics. Their doctrine, Stoicism, remained an important philosophical school well into the Roman era, when their leading thinkers were the Roman emperor Marcus Aurelius (AD 121–180), the writer Seneca (c. 2 BC–AD 65), and the philosopher Epictetus (c. AD 55–c. 135). In its later form, Stoicism reflected an increasing concern with ethical and moral questions.

Side by side with the philosophers, Greece also produced men who had practical concerns, who created or perfected tools and machines. One of the greatest among these creators was the inventor and mathematician Archimedes, who was born in 287 BC. He lived in Syracuse, a Greek colony in Sicily, and left many treatises in arithmetic, mechanics, astronomy and optics. Archimedes is best known for the machines that he built, like catapults and other military engines. He is also said to have proposed the use of concave mirrors to set the Roman navy on fire. He was killed as the Romans conquered and sacked Syracuse in 212 BC.

Magnetism in Greece

There are but few references to magnetism in the writings of the early Greek philosophers; the comment on the magnet attributed to Thales, quoted at the beginning of this chapter, is the first ever recorded. Some philosophers, besides their considerations on other natural phenomena, not only described magnetism, but also attempted to explain the cause that lay behind the bizarre behavior of the lodestone. A statement to this effect was reported by Aristotle, in his treatise *De Anima* ('On the Soul'): "Thales, too, to judge from what is recorded about him, seems to have held the soul to be a motive force, since he said that the magnet has a soul in it because it moves the iron".[48]

In Diogenes Laertius' *Lives of the Philosophers* (early 3rd century AD), the same idea is attributed to Thales[49]: "Aristotle and Hipias say that he ascribed souls to lifeless things too, taking the magnet and amber as his evidence." This idea attributed to Thales has been interpreted either as meaning that Thales shared the traditional beliefs of his contemporaries, or that he had altogether abandoned them; the latter view is based on the fact that he had reserved this identification with spiritual beings only for admittedly abnormal objects, like the magnets[50].

Laertius lists amber side by side with the magnet, as notable materials that are the object of Thales' remark. Amber is fossil tree resin, of yellow to orange color, transparent or milky, and is known for the property of attracting small objects when rubbed, due to the appearance of an electric charge. Amber is also mentioned by Pliny the Elder; in his words[51], "it attracts straw, dry leaves and bark from the linden tree, just as a magnet attracts iron" (see Chapter 3).

In the Dialogues of Plato, one finds a description of how a magnet makes a piece of iron it touches behave as a magnet to another piece of iron, a phenomenon usually known as induction[52]: "(...); there is a divinity moving you, like that contained in the stone which Euripides calls a magnet, but which is commonly known as the stone of Heraclea. This stone not only attracts iron rings, but also imparts to them a similar power of attracting other rings; and sometimes you may see a number of pieces of iron and rings suspended from one another so as to form quite a long chain: and all of them derive their power of suspension from the original stone."

The Greek philosophers after Thales also tried to explain why the magnet had the power to attract iron. The philosopher Empedocles (c. 490-c. 430 BC) was one of them; he appealed to 'effluences', or vapors, in his explanation. According to the book *Quaestiones*, written by a philosopher who was active around AD 200, Alexander of Aphrodisias, this was Empedocles' idea for explaining magnetic attraction:

"On the reason why the lodestone attracts iron. Empedocles says that the iron is attracted to the stone by the effluences which issue from both, and because the pores of the stone are commensurate with the effluences from the iron. The effluences from the stone stir and disperse the air lying upon and obstructing the pores of the iron and when this is removed the iron is drawn on by a concerted outflow. As the effluences from the iron travel towards the pores of the stone, because they are commensurate with them and fit into them the iron itself follows and moves together with them."[53]

Along the same line, Democritus of Abdera (c. 460-c. 370 BC), a Greek philosopher best known for his atomistic view of matter, also appeals to effluences to explain the properties of the magnet. In the words of Alexander of Aphrodisias: "Democritus also says that there are effluences and that like bodies are attracted to like, but adds that all are attracted to a void. Having made these hypotheses, he supposes that the lodestone and iron consist of similar atoms, but those of the stone are smaller and it is of rarer texture than the iron and contains more void. For this reason, its atoms being more mobile are attracted more quickly to the iron (for they are moving to their similars), and entering the pores of iron disturb the atoms in it as they pass between owing to their small size. The atoms of the iron, thus disturbed, stream outside towards the stone because of their similarity and because it has more void. The iron [as a whole] follows them in their wholesale expulsion and movement and is itself drawn towards the stone. The reason why the stone does not move any more towards the iron is that the iron does not contain so much void." [54]

Epicurus of Samos (341–270 BC), the philosopher known for his teachings that pleasure, or rather absence of pain, was the essence of life, had a different theory; according to the physician and philosopher Galen (c. AD 130-c.200)[55], "His view is that the atoms which flow from the stone are related in shape to those flowing from the iron, and so they become easily interlocked with one another; (...)".

In every case, the proposed explanations involve mechanical actions, attraction by the void, interlocking of atoms, and so on, (or maybe also animistic considerations, in the case of Thales with his remark on the "soul of the magnet"). Mechanical processes were certainly insufficient to explain magnetic effects; students of magnetism would have to wait for more than 20 centuries for a sound explanation of the phenomenon.

Chinese Records on Magnetism

Some of the earliest references to the lodestone come from the East, more specifically from China, where it was called 'tzhu shih' – the loving stone[56]. The first practical application of magnets – the compass – also originated in China, where it may have been used for navigation in the 10th century AD.

The first Chinese dynasty was the Shang dynasty, in the period from the 18th to 12th centuries BC, in the Chinese Bronze Age. Inscriptions found in pottery point to the origin of Chinese written language as early as 4000 BC. The unification of China did not occur until 221 BC, as the powerful feudal state of Ch'in established its dominance over the other states; in the process of unification, the written language was standardized.

During Chinese antiquity, there were enormous advances in the knowledge of natural phenomena, and in different technologies. Astronomical knowledge reached a very high level, mostly stimulated by the need to perfect the calendar, and also for divination purposes. By 1400 BC, the Chinese had already determined the duration of the year as $365\frac{1}{4}$ days (the accepted value today is 365.242199 days), and the period of the lunar cycle of phases (called synodic month, or lunation) as $29\frac{1}{2}$ days (known today to be 29.530588 days). They had a 354-day year, and from time-to-time added an extra month with 29 or 30 days to keep the calendar in line with the motion of the Earth around the sun. They kept registers of astronomical phenomena that are invaluable today for the study of past astronomical events: they had recorded eclipses since 720 BC, sunspots since 28 BC, comets since 613 BC, and novae and supernovae since 352 BC[57]. The collection and preservation of these records can in general be attributed to the activity of the state, and this strong presence of the state is a distinguishing mark of the early Chinese scientific development[58].

The Chinese started to make objects of cast iron as early as the 6th or 4th century BC, thanks to the invention of piston bellows that provided a steady flow of air into the smelting furnace. Among other technological breakthroughs, one may include paper, invented in China around AD 100, a mechanical clock in the 8th century, and gunpowder in the 9th century. In many technical fields, the Chinese led Europe until the 16th century.

The first important thinkers in ancient China were the Confucians and the Taoists, but there were also relevant contributions from members of other philosophical schools, known as Mohists, Legalists and Logicians. The Confucians were the followers of K'ung-fu-tzu, or in Latinized form, Confucius, who lived from 552 BC to 479 BC. The Taoists (from Tao, 'The Way'), followed the teachings of Lao-tzu, said to have lived some time between the 6th and 4th centuries BC, and the Mohists were followers of Mo-tzu (479–381 BC). Although Confucians and Taoists were the most influential schools in Chinese intellectual history, and their teachings eventually evolved into two religions, Mohists, Logicians and Legalists were more immediately related to the development of scientific ideas in China. The Mohists studied optics and mechanics, and developed, together with the Logicians, the fundamentals of a scientific logic[59]. They discussed the different forms of acquisition of knowledge: through hearsay, by inference, by direct observation, and by deliberate action, that is, by experimentation[60]. The Legalists, who flourished around 300 BC, preached in favor of laws that would strictly determine human conduct, and came close to the concept of laws of nature. They also contributed to an incipient scientific attitude with their preoccupation with measure and quantification[61].

According to an anonymous manuscript at the time of the Han dynasty (206 or 202 BC to AD 220), Chinese mathematical knowledge included the computation of areas of geometric figures, volume of prisms, pyramids, cylinders, the solution of equations of the first degree, solution of systems of linear equations, and application of Pythagoras' theorem.

The Chinese studied the properties of magnets from a very early date, and also developed the first applications of magnetism*. The first written record on the magnet is found in the *Lü Shih Chhun Chhiu* ('Master Lü's Spring and Autumn Annals'), written by a group of scholars and published in 240 BC, in the Chou period. It describes the property of the lodestone and asserts[62]: "The lodestone draws to itself iron particles." Other texts contain further observations on the lodestone: for instance, that it does not attract other metals or non-metallic objects[63]. As in this quotation from *Huai Nan Tzu* ('The Book of the Princes of Huai Nan'), written before 120 BC: "If you think that be-

* This section relied heavily on J. Needham's monumental study *Science and Civilisation in China*, see Further reading and Ref 56.

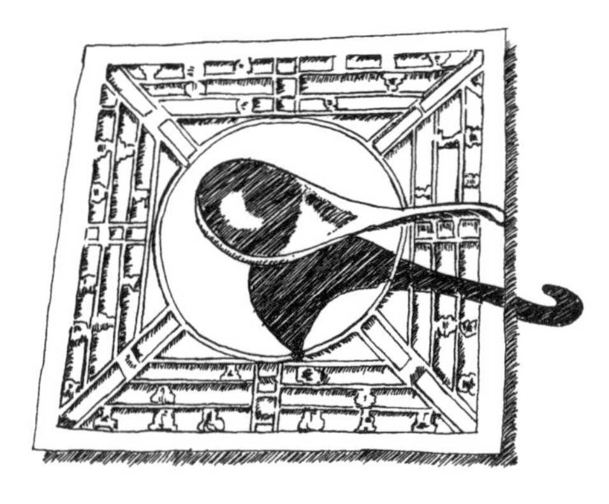

Figure 1.3 Chinese divination board with "south-pointing spoon" made of magnetite.

cause the lodestone can attract iron you can also make it attract pieces of pottery, you will find yourself mistaken. (...) Fire is obtained from the sun by the burning-mirror, the lodestone attracts iron, crabs spoil lacquer, the mallow [malvaceous plant] turns its face to the sun. Such effects are very hard to understand." Also, from the same book: "Some effects are more pronounced at short range and others at long range. Rice grows in water but not in running water. The purple fungus grows on mountains, but not in stony valleys. The lodestone can attract iron but has no effect on copper. Such is the motion (of the Tao)."[64] And from the *Lun Hêng* ('Discourses Weighed in the Balance'), book of AD 83: "Amber picks up mustard-seeds and the lodestone attracts needles."[65]

In the 3rd century BC, the Chinese used a diviner's board for fortune telling. It consisted of two parts, the upper one (circular) representing the heavens, resting on a square board representing the Earth, with divisions corresponding to the compass points. The diviner threw small stones or figures onto the lower part, and told the future from the positions they occupied. Among the figures used, one represented the constellation of the Great Bear (The Northern Dipper), in the shape of a spoon, made of wood, pottery or stone. In the 1st century AD, (and possibly even earlier, in the 2nd century BC) the spoon was made of lodestone, and from its behavior, became known as the 'south-pointing spoon' (Figure 1.3). This was the first example of an application of the property of a magnet of orienting itself in the Earth's magnetic field. A remark on the property of this spoon is found in the *Lun Hêng*[66]: "But when the south-controlling spoon is thrown upon the ground, it comes to rest pointing at the south." Chinese texts also mention a cart with a figure mounted on it that always pointed to the same direction. According to Needham, this had nothing to do with magnetism; it was a purely mechanical device[67].

In the 1st century AD lodestones mounted on a pin, which allowed them to turn more freely, were common; in the 7th or 8th century, iron needles magnetized through contact with the lodestone substituted the pieces of rock, becoming the first needle compasses. The application of this early compass to navigation may have occurred in the 10th century, but the instrument was certainly being used by the 11th century. The first Chinese description of a magnetic needle compass is from around AD 1080, 100 years before the first European written records of such devices[68]. One of the first Europeans to discuss the use of the compass was the Englishman Alexander Neckam (1157–1217), in his book *De Nominibus Utensilium* ('On the Names of the Tools'), written around 1175–1183.

Magnets in Pre-Columbian Mesoamerica

The first complex culture that appeared in the Americas was that of the Olmecs, in about 1200 BC, centered in the coastal areas of Mexico, in the present states of Vera Cruz and Tabasco. The main sites are along the Gulf of Mexico: San Lorenzo, La Venta, Laguna de los Cerros and Tres Zapotes. The Olmec culture attained a high degree of sophistication, with a knowledge of building techniques and the manufacture of implements. They influenced the later civilizations that developed on the continent, especially the Aztecs and the Mayas.

The Olmecs left impressive archeological remains, the most remarkable being their characteristic colossal stone heads, weighing over 15 tons; some sixteen of these have been found. The Olmec society was able to mobilize manpower on a large scale for building and earth-moving work. The stone for some of the heads was quarried some 80 km away, in the Tuxla mountains, floated down a river and dragged overland, a feat in itself.

Archeologists have found Olmec objects made of iron ore, dating from the Early Formative period (1500–900 BC) in sites in San José Mogote and San Lorenzo[69]. These are small polished plates or mirrors, plane or concave, up to 10 cm (4 inch) in diameter, which were apparently used as body adornments[70]. It is quite possible that the manufacture of the mirrors made of magnetite could lead to the observation of magnetic behavior, for example, the adherence of the residues resulting from the process of cutting and polishing the objects.

One of the most interesting Olmec objects, found in strata dated 1400–1000 BC, was a polished bar of 3.5 cm (1 1/3 inch) that is magnetic, and has led to speculations that this was the earliest known compass[71]. If true, this would mean that the Olmecs had anticipated the Chinese by more than one thousand years.

Further evidence that the Olmecs knew about the properties of magnetic ores was the discovery in the coastal plain of Guatemala of a statue of a jaguar with magnetic poles in each raised paw, and a crude statue of two seated men made of a single block of stone, with magnetic poles on either side of the navel. The latter statue is dated from 2000–1500 BC, therefore it would represent the oldest known magnetic artifact in the world[72]. In Izapa, in a site corresponding to the Late Formative period (300 BC-AD 100), a carved stone turtlehead of 1.1 × 1.2 m (45 × 48 inch) that is also magnetic was found, with one of the magnetic poles coincident with the snout of the animal[73]. Although the magnetization through the impact of lightning is a possibility, it seems unlikely in view of the significance of the location of the magnetic poles in the different objects.

Roman Sources

The expansion of Rome in the 3rd and 2nd centuries BC led to the Roman occupation of the Greek towns in the South of the Italian peninsula, after the Pyrrhic War (280–275 BC), and later, after the Macedonian Wars (214–148 BC), assured complete control of Macedonia and mainland Greece. The Romans incorporated the achievements of Greek culture in the arts, in philosophy and in science. However, no blossoming of creative ideas to match the Greek phenomenon is recorded. A representative of the scientific tradition of the Roman world was the philosopher Lucretius (c.98-c.55 BC), already mentioned. He described in his book *De Rerum Natura* ('On the Nature of Things') some properties of the magnet, and how its power can be felt through a metallic obstacle: "I have seen Samothracian iron rings even jump up, and at the same time filings of iron rave within brass basins, when this Magnet stone had been placed under; such a strong desire the iron seems to have to fly from the stone."[74]

Another important Roman author was Pliny the Elder (AD 23–79), also quoted above. He was a writer and an investigator who left several treatises, the most famous of all being the Natural History. Pliny showed a deep interest in the magnet, for its "striking properties": "For what phenomenon is more astonishing? Where has nature shown greater audacity?"[75] He describes the types of magnetic stones, distinguishing them as 'male', or 'female', according to the strength of their magnetic effects. He even speaks of a mythical stone from Ethiopia – *theamedes* – that "repels every kind of iron"[76]. He tells tales of two mountains near the River Indus, one that attracts iron, other which repels it. Therefore, if a man has nails in his shoes, on one mountain "at each step he is unable to tear his foot away from the ground and on the other he cannot put his foot down[77]".

The only way one can observe repulsion with a magnet, of course, is by approaching another magnet, with North pole near North pole, for example. There appear to be no reports of interaction between magnets in the ancient texts, and consequently no genuine observation of magnetic repulsion[78].

"There is a stone, colorless and without brilliance, (...) but it is preferred to all the most precious products of the Orient by those who know it virtues and its wonders." Thus wrote another Roman author, the fourth-century poet Claudian, or Claudius Claudianus (c. 370-c. 404) in a text[79] that extolled the magic qualities of the magnet. He continued: "Iron gives it life; iron recognizes and nourishes it; when the iron is withdrawn, it experiences the torments of hunger and thirst; it dies (...) Iron and the loadstone are drawn together and united. What can be this subtil flame which, entering these two metals, can give rise to this sympathy? What is the unknown charm which can unite them with a common will and a single desire?"

When Greek science entered a declining phase after the 2nd century AD, its tradition was continued by the Arabs, whose scientific thought was strongly patterned after the Greeks[80]. Their scientific contribution will be discussed in Chapter 2.

The next great step in the development of scientific explanations, after the Greeks had laid the groundwork, required a new critical attitude, a scientific revolution that enthroned the experimental method and the criticism of the classics as prime tools in the search for knowledge, in the 16th and 17th centuries. A revolution that brought with it a new view of the universe, with the Earth at the same time integrated into it, and removed from its privileged position.

Further reading

J. Barnes, *Early Greek Philosophy*, Penguin, London, 1987.

M. R. Cohen and I. E. Drabkin, *A Source Book in Greek Science*, Harvard University Press, Cambridge, 1966.

W.K.C. Guthrie, *A History of Greek Philosophy*, vol. I, Cambridge University Press, Cambridge, 1967.

J. Needham, *Science and Civilisation in China*, vol. 4, part I, Cambridge University Press, Cambridge, 1972.

J.M. Roberts, *The Penguin History of the World*, 3rd edition, Penguin Books, London, 1997.

C.A. Ronan, *The Cambridge Illustrated History of the World's Science*, Cambridge University Press, Cambridge, 1984.

Chapter 2
The Finger of God

"this stone bears in itself the likeness of the heavens"

Letter of Peter Peregrinus de Marincourt, *August 8, 1269*

Introduction

The Middle Ages begin with the decline and subsequent fall of Rome to the invading Visigoth armies, in the 5th century, and last until the fall of Constantinople in the 15th century. During the Middle Ages, the relevant sources of scientific knowledge were Latin Europe, the Greek world, China, India, the Arab world, and pre-Columbian America. Only in Western Europe, however, would the scientific enterprise evolve into a new paradigm in the 17th century, a transformation generally described as the scientific revolution.

From the point of view of scientific development in the West, the Middle Ages can be divided, in a schematic way, into four different periods[1]: 1) the dark age, or high Middle Ages (5th–10th centuries); 2) the awakening of Europe and the Islamic influences (11th–12th centuries); 3) the flourishing of the universities and the golden age of 'scholastic' science (13th and beginning of the 14th century) and finally, 4) the interdependence of the sciences and technologies (AD 1350–1450).

The advances in the sciences and technology in the Middle Ages were very significant[2], comprising in the first place, a change in the perspective of science, with the growth of the idea of control of nature by man, and an attitude of disapproval toward restricting man's speculation to the limits of one single scientific or philosophical system. The Middle Ages also witnessed the renaissance of the Greek idea of theoretical explanation in the sciences, and the gradual shift in interest from metaphysical questions of causes to the mathematical description of phenomena.

From Lodestone to Supermagnets. Alberto P. Guimarães

ISBN: 3-527-40557-7

In the physical sciences, notable progress stems from the first attempts (in the 14th century) to formulate a theory of motion. Among the important technological developments, new machines appear, devised to exploit waterpower and animal power; the mechanical clock and magnifying lenses; the astrolabe and the quadrant. Also worth noting are the advances in medicine and surgery, in the descriptions of diseases and descriptions of fauna and flora of different regions.

A general trait to be found in the scientific thought of that period is the fact that questions posed by medieval scholars were frequently mingled with questions of philosophy of science[3]; this remained true of science as practiced up to the 17th century, and reflected the search for a new scientific paradigm.

Arab and Chinese Sources

The Arabs started to expand their empire out of the Arabian Peninsula in the 7th century AD. At its pinnacle, in the middle of the 8th century AD, the empire extended from Spain to India. The capital of the empire was Baghdad, founded in AD 762 by the caliph al-Mansur (AD 709/714–775) of the Abbasid dynasty, on the site of an early Persian village. Baghdad reached a very high level of development in the succeeding years, when it reached half a million inhabitants, coming to be regarded the richest city in the world in the period from the 8th to the 9th century.

In the 7th century, the Prophet Muhammad (c.AD 570–632) founded Islam as a new religion, with one omnipotent and omniscient God (Allah). The teachings of the Islamic sacred book, the *Qur'am*, (or Koran, which means in Arabic: Reading or Recitation), said to be revealed by God to Muhammad, valued medicine and knowledge in general, a factor that helped the development of the sciences in the Arab world. Arab scholars were particularly fruitful in their research in astronomy, producing some very precise tables that surveyed the positions of stars and planets. They developed mathematics, especially trigonometry, and algebra, having created the word, from the Arabic *al-jabr*, meaning, 'to restore', from the balancing of the two sides of an equation. They were responsible for the import, in the 7th century, of Hindu numerals, that were later universally adopted, and are the symbols used today, known under the name of Arabic numerals. They were also the first to use the zero in the positional notation, as used today.

The original contributions of the Arabs are sometimes belittled in comparison with their important role in preserving and making available to Western Middle Age thinkers the works of the Greek philosophers and scientists. This effort to save the older manuscripts in fact started long before the flourishing of Arab sciences, when scholars took steps to preserve valuable texts before the destruction of the library and museum in Alexandria in the 5th century. Later, in important Arab centers of learning, especially in Baghdad and Cordoba, systematic programs of translation of the Greek works were carried out. Arab scholars not only translated the classic texts, but also made many additions and commentaries that very much enhanced their value. Under the Arabs, science had for the first time an international character, thanks to the extension of the kingdom in the mid-eighth century, from the Iberian Peninsula to the Indian subcontinent, and favored by the unifying forces of common religion and language[4].

The greatest contribution to physics by Arab scholars was the work of ibn al-Haytham (known in the West as Alhazen) (c. 965–1039), born in Basra, Iraq. He worked in the library in Cairo, during the reign of the caliph al-Hakin. He wrote a treatise, called 'Treasury of Optics' (*Kitab al-manazer*), and made important steps toward the understanding of optical phenomena. He rejected the traditional Greek picture of light emitted by the eyes, used the concept of 'light rays' that propagated in straight lines, and correctly interpreted that refraction of light in the boundary between two transparent media was due to the difference in velocity of propagation in the two media[5].

References to magnetism are not very common among Arab authors; the first text on the magnetic compass by an Arab author (writing in Persian) contained the stories compiled by Muhammad al-Awfa (d. c. 1230), in the *Jami al-Hikayat*, published around 1232. In 1282, Bailak al-Qabajaqa, of Cairo, was the first to write in Arabic on the use of the magnetic compass for navigation. In the treatise of Tayfashi (13th century) the existence of the two poles of the magnet is reported, and also its property of indicating direction[6]. Arab mariners may have used the compass as early as in the 11th century[7].

In China, the existence of magnetic poles, and the fact that the magnetic North did not coincide with the geographical North, were known by the 8th or 9th century. The latter discrepancy would not be discovered in the West for another seven centuries. This difference in direction between the geographical North-South line (the meridian) and that of the compass is called magnetic declination. It arises from the fact that, although the compass needle points roughly in the North-South direction, the Earth's North and South magnetic poles are not precisely located on the corresponding geographical poles, which are the points that mark the axis of rotation of the planet on the Earth's surface. The position of the magnetic poles on the surface of the planet has varied throughout geological time, and in fact, North and South magnetic poles have reversed several times, on the average about once every 280 000 years in the last six million years[8]. This effect is not completely understood; we will return to the subject of terrestrial magnetism in Chapter 6.

The Chinese treatise on military technology *Wu Ching Tsung Yao* ('Collection of the Most Important Military Techniques') in AD 1044 contained a description of the preparation of a compass with a magnetic floating needle that did not require contact with the lodestone to be magnetized. It reads[9]: "Now the carriage method has not been handed down, but in the fish method a thin leaf of iron is cut into the shape of a fish two inches long and half an inch broad, having a pointed head and tail. This is then heated in a charcoal fire, and when it has become thoroughly red-hot, it is taken out by the head with iron tongs and placed so that is tail points due north. In this position, it is quenched with water in a basin, so that its tail is submerged for several tenths of an inch. It is then kept in a tightly closed box. To use it, a small bowl filled with water is set up in a windless place, and the fish is laid as flat as possible upon the water surface so that it floats, whereupon its head will point south."

This method of magnetization of the needle relies on the fact that a piece of iron cooled in a magnetic field – in this case the Earth's field – retains a certain magnetization, called remanent magnetization. This occurs because the majority of the small magnetic regions in the metal (called domains), formed as the metal cools, are oriented in the direction of the applied field, retaining a net magnetization (this will be discussed in more detail in Chapter 5).

Magnetism and the Compass in Europe

The diffusion of scientific knowledge in Europe during the early Middle Ages owed much to works of encyclopedic character, such as Pliny's Natural History, the *Etymologies* of Isidore of Seville (560–636) and the *Geometry of Boethius*, by Anicius Manlius Severinus Boethius (c. 480-c. 525). These works, although very valuable, are mostly compilations of earlier knowledge. They were followed by the works of Venerable Bede (673–735), Alcuin of York (735–804) and Hrabanus Maurus (776–856), each one transcribing much from their predecessors. At the end of the 10th century, a noteworthy scholar in the Christian West was Gerbert of Aurillac (c. 945–1003), who wrote the logical treatise *De rationali et de ratione uti* ('Concerning the Rational and the Use of Reason') and became the first French pope, as Sylvester II. He had contacted Arab scholars during a stay in Spain, and after that helped to propagate the use of the Arabic numerals.

After the 6th century, when St. Benedict of Nursia (or Norcia) (c. 480-c. 547) founded the monastery of Monte Cassino, in Italy, the preservation and diffusion of scientific knowledge took place in the monasteries and in the adjoining schools. The monks of the order, known as the Benedictine, maintained large libraries, and systematically copied religious, philosophical, and some literary works. In the 8th and 9th centuries, schools were created at the cathedrals, for example in York in England, and Orléans, in France. They were initially dedicated to the education of priests, but later accepted lay students. Starting from the 11th century, when the University of Bologna was created, until the 15th century, a movement of creation of centers of higher learning swept over Europe: they were founded in Oxford (1167), Paris (1170), Salamanca (1218), Padua (1222), Cambridge (1290), Rome (1303), Florence (1321), Vienna (1365), Heidelberg (1386), Saint Andrews (1411), and Uppsala (1477), among others. These new centers were to become important nuclei of scholarship in the sciences.

Many Western authors of the early Middle Ages described the properties, real or legendary, of the magnet. Among these, one can name the Latin grammarian Caius Julius Solinus (2nd and 3rd century AD), the theologians Augustine (354–430) and Isidore of Seville[10], already mentioned. Later, the Bishop of Rennes, Marbode (1035–1122) in the *Liber de Gemmarum* ('Book on the Gems'), helped to propagate many legends related to the uses of the lodestone; he referred to the employment of a magnet to expose unfaithful wives – by placing the magnet under their pillow, they will fall from the bed if the suspicion is founded. Also, magnets were said to be employed by thieves to drive out the occupants of a house[11]; one can destroy the power of the magnet by rubbing its surface with garlic, or goat's blood. Many of these uses of the magnet were taken from the *Historia Naturalis*, of Pliny the Elder; the reference to garlic seems to have arisen from an error of translation.

While the application of magnetism to the construction of the magnetic needle compass was described in China in a publication around AD 1080, in the West the first references to the compass appeared about one century later. The available evidence leads one to think that the compass evolved independently in the East and the West*.

Three authors are nowadays regarded as pioneers in reporting the use of the magnetic compass in the West: these are the Englishman Alexander Neckam (1157–1217), and the Frenchmen Guyot de Provins (fl. 1184–1210) and Jacques de Vitry (c. 1165–1240), in this order of chronological priority**.

Alexander Neckam was born in St. Albans, some 20 miles (32 km) northwest of London, in 1157, and lectured in Paris and Oxford; he was appointed abbot of Cirencester in 1213. Neckam is the author of several books, among them *De Nominibus Utensilium* ('On the Names of the Instruments') and *De Naturis Rerum* ('On the Nature of Things'), both containing mentions of the magnetic compass. The earliest reference is to be found in *De Nominibus Utensilium*, written around 1175–1183[12]: in a ship, there should be "a needle mounted on a pivot, which will oscillate until the point looks to the east [sic], and the sailors will know how to direct their course when the northern constellation of the Little Bear is obscured by the troubled [state of the] atmosphere; for it never disappears below the horizon, because of its small circle." A second mention by Neckam is given in his book *De Naturis Rerum*, of 1197–1204: "So sailors crossing the sea, when because of overcast skies they lose the Sun's light, or when the world is wrapped in the darkness of night, and they do not know what cardinal point the ship

* The basic reference used here on the work of Peregrinus and his predecessors is J. A. Smith (Ref 10).

** J. A. Smith (Ref 10) argues against the dates given by other science historians.

is headed toward, put a needle above the lodestone; and the needle revolves until, after its motion has stopped, its point faces due north"[13].

Neckam recounts the old legend of the iron statue of Muhammad suspended in mid-air through the action of magnets; the same tales of magnets sustaining statues also appear in Pliny's *Historia Naturalis.* A similar legend tells that in India, the Sun temple of Konarak, in the state of Orissa, had a statue held from a large lodestone mounted under the vault[14].

Guyot de Provins, a cleric and minstrel at the court of the Emperor Frederick I, or Barbarossa, wrote *La Bible,* a satirical poem published probably in 1206, where he makes a metaphor involving the magnetic compass, and proceeds to describe its use by the mariners. The other claimant to priority is the historian and cleric Jacques de Vitry, who refers to the compass in his book *Historia Orientalis Hierosolymitana* ('Eastern History of Jerusalem'), a text on which he worked for many years; the part on the compass seems to have been written in 1204.

These references treat the magnetic compass as an instrument already known for some time; from this fact, one is led to date tentatively the appearance of the compass in Europe at about AD 1150[15]. Compasses mounted on pivots ('dry' mounting) and compasses floating on water ('wet' mounting) are equally present in these early references, which suggests that these two forms of compasses appeared in Europe at approximately the same time.

Early medieval authors often used the word adamant, from the Greek *adamas,* 'invincible', meaning either the lodestone or a very hard material, sometimes diamond, and in some other passages, steel. The source of confusion between these meanings is the incorrect attribution of the origin of the word to the Latin *adamare,* or 'having attraction for'. Therefore *lapide adamanten* became the attracting stone, from which derived in the Romance languages aimant (French), imán (Spanish) and ímã (Portuguese)[16].

The French philosopher and theologian Guillaume d'Auvergne (also referred to as William of Auvergne, or William of Paris) (c. 1180–1249), a follower of Augustine, wrote *Magisterium Divinale* ('The Divine Teaching'), his major work, in the period 1223–1240. Guillaume d'Auvergne was a professor in Paris, appointed around 1225, and became the bishop of Paris in 1228. In his work *De Universo Creaturarem* ('On the Universe of Created Things'), he discusses the way a magnet attracts one piece of iron, and then another piece in contact with the first piece, the phenomenon of 'induction' (see Chapter 1). He is one of the first authors to suggest that magnetic forces are also active in the sky; he uses the idea of induction to explain how the motion is transferred, at a distance, from one heavenly sphere to the other[17]: "Then what is so astonishing if the virtue of life or of the soul of the first celestial sphere transmits itself to the second, then the second to the third, and so forth until it reaches the last moving sphere, which is the sphere of the moon? That cannot itself produce the same if there is not, between them, another connection between closeness and contact, as happens in the proposed example [i. e. the magnet]."

The Belgian physician John of St. Amand (1261–1298) was a canon of Tournai, in the Southwest of Belgium; among the many books he wrote, the *Expositio sive Additio Super Antidotarium Nicolai* ('A Commentary on the Antidotary of Nicholas of Salerno') contains passages that anticipate the idea that the Earth is a magnet, an idea that was developed later by the Englishman William Gilbert, at the end of the 16th century. He states[18, 19]: "Wherefore I say that in the magnet is a trace of the world, wherefore there is in it one part having in itself the property of the west, another of the east, another of the south, another of the north."

Magnetic Epistle from the Trenches

Charles of Anjou (Charles I) (1226–1285) and his brother, the French King Louis IX, participated in the ill-fated Crusade in Egypt in the years 1248–50. In the 1260s, Charles led a military campaign in Sicily and in mainland Italy, conquering the island and Naples, defeating King Manfred (c. 1232–1266?), of the German Hohenstaufen dynasty, then in power. In 1269, he held under siege and took Lucera, in Apulea (Puglia), near the city of Foggia, in the heel of the Italian boot. In the trenches encircling Lucera, in that year, the first European work entirely devoted to the discussion of the magnetic properties of the lodestone, the *Epistola de Magnete* ('Letter on the Magnet') was written* (Figure 2.1). It was the work of the Frenchman Pierre de Marincourt

* It is not known when the text was written, only that it was concluded in the year 1269.

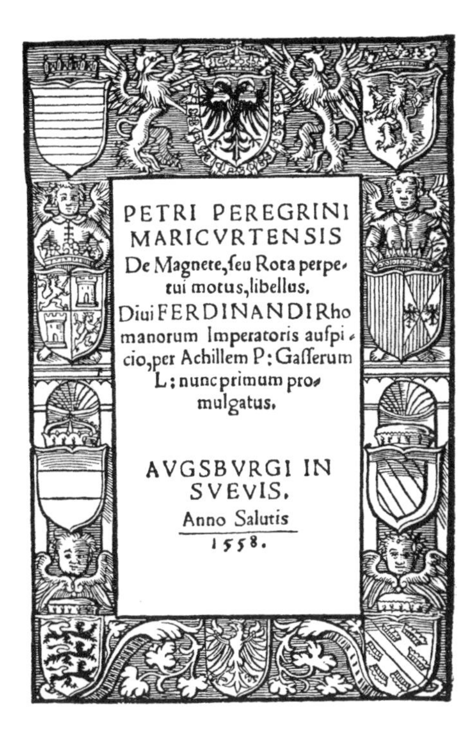

PETRI PEREGRINI
MARICVRTENSIS
De Magnete, ſeu Rota perpe-
tui motus, libellus.
Diui FERDINANDI Rho
manorum Imperatoris auſpi-
cio, per Achillem P: Gaſſerum
L: nunc primum pro-
mulgatus.

AVGSBVRGI IN
SVEVIS.
Anno Salutis
1558.

Figure 2.1 Title page of the Epistola de Magnete (Letter on the magnet) (1269), by the Frenchman Peter Peregrinus (b. c. 1220), in an edition of 1558 (courtesy Bakken Museum).

(b. c. 1220), also known by his Latin name, Petrus Peregrinus[20]. He may have received the title of 'Peregrinus' not for visiting the Holy Land or participating in the Crusades, as was usual, but for his role in that military campaign. Pope Clement IV (d. 1268), also a Frenchman, had declared the attacks against the Hohenstaufens as official Crusades, since the Germans had the support of the Saracens, the name by which the followers of Islam where known at the time.

The work of Peregrinus was written in the form of a letter to a soldier, a certain Siger, of Foucaucourt, in Picardy. In this short letter, written in response to questions put by the same Siger, Pierre de Marincourt relates some properties of the magnet, describes experiments with it, and mentions its applications. The influence of the *Epistola* was very large: it was cited by many books, such as William Gilbert's *De Magnete* ('On the Magnet') (1600) and Athanasius Kircher's (c. 1602–1680) *Magnes, sives De Arte Magnetica* ('The Magnet, or About the Magnetic Art') (1641), and plagiarized in *De Natura Magnetis* ('On the Nature of the Magnet') (1562) of Jean Taisnier[21]. There are at least 31 known manuscript versions of the *Epistola*[22].

Little is known about Pierre de Marincourt. At the time of the letter he was most likely an engineer in the army; he probably came from the town of Méharicourt, in Picardy, a province corresponding roughly to the present French *région* of the same name, in the North of the country. He seemed to be primarily interested in designing and building scientific instruments; he declares the *Epistola* "part of a work on the construction of philosophical instruments"[23]. He also mentions another book on the action of mirrors, which he was either writing or intended to write.

"The disclosing of the hidden properties of this stone is like the art of the sculptor by which he brings figures and seals into existence", Pierre de Marincourt writes in the first part of the work. He then continues to list the physical properties that characterize the specimens of lodestone: the color, homogeneity, weight and strength; by strength he means the power to attract pieces of iron.

He describes the two regions of the lodestone which reveal a stronger magnetic attraction as "poles", and chose this denomination in analogy to the celestial sphere[24]: "I wish to inform you that this stone bears in itself the likeness of the heavens, as I will now clearly demonstrate. There are in the heavens two points more important than all others, because on them, as on pivots, the celestial sphere revolves: these points are called, one the arctic or north pole, the other the antarctic or south pole." He refers to the celestial poles, not to the Pole star, since he knew that the Pole star does not sit exactly on the celestial North Pole. Peregrinus' association of the properties of the lodestone to the heavens was in line with scholastic philosophy, or more specifically with the thought of his contemporary St. Thomas Aquinas (1224/25–1274), the greatest philosopher of the Christian Middle Ages, who also attributed the effects of the magnet to *virtus coeli*, the "power of heavens"[25].

Peregrinus discusses how a lodestone, broken into two pieces, shows two pairs of poles, how a piece of iron becomes magnetic after touching a lodestone, how like poles are repelled and opposite poles attracted. Although earlier authors had already described magnetic repulsion, the association of repulsion to equal poles is made here for the first time.

The property of the lodestone of aligning in the North-South direction was already known; this is a very remarkable property, of course, impossible to predict with the knowledge of natural phenomena then

Figure 2.2 Compass needle mounted on a pivot ("dry mount").

available. Why would the stones that attracted pieces of iron have anything to do with the cardinal points on the Earth's surface? When one considers closely this extraordinary fact, one understands why the British physicist John Desmond Bernal (1901–1971) regarded this as the greatest discovery in the history of physics![26] Equipped with the knowledge of this property, Peregrinus then proceeds to show how to make a floating compass with a scale of 360°: this instrument is regarded as the first compass with the proper scale division[27]. He also described, besides this 'wet' compass, a 'dry' one, i.e. one where the needle turns on pivots (Figure 2.2).

Peregrinus was the first to shape a lodestone; he cut it as a sphere, but stopped short of making the analogy of the spherical magnet with the planet Earth. Since the magnet seemed to be linked to the heavens, he believed (erroneously) that a spherical lodestone pointed to the pole, if free to turn, would rotate continuously, following the motion of the stars in the sky. Such a lodestone would then make a complete turn in 24 hours, and could be used as a clock.

He rejected the idea that the lodestone tended to point to the poles due to the presence of magnetic rocks in the Earth's poles. He argued that if this were true, since there are iron mines in many parts of the world, the compass would point to completely different directions at different places.

In the last part of his Epistle, Peregrinus gives the design for a perpetual motion machine based on magnetic attraction and repulsion. To justify his proposal, he affirms[28]: "I have seen many persons vainly busy themselves and even becoming exhausted with much labor in their endeavors to invent such a wheel. But these invariably failed to notice that by means of the virtue or power of the lodestone all difficulty can be overcome." In his project he had the teeth of an iron cogwheel attracted by a lodestone; as the lodestone approached each tooth, it turned an axle. This model of a perpetual motion machine, although condemned to failure like all the others, may have been the first attempt to design such a machine based on scientific premises[29].

In his *Opus Tertium* ('Third Work'), written in 1267, the English philosopher Roger Bacon (1214–1294) refers with words of praise to an investigator who is "a master of experiments (*dominus experimentorum*)". "I know of only one person who deserves praise in the works of this science", he adds in another passage, speaking of the use of burning mirrors[30]. In annotations made in the margin of the text, this person was identified as Pierre de Marincourt; however, the authenticity of this attribution has not been confirmed, since these comments had possibly been added by a scribe[31]. In these notes, the work of Peregrinus is also valued for his method, especially for the importance he attributed to experiments.

The main importance of the Epistle resides in the fact that it provided the first printed systematization of the phenomena of magnetism, and that its contents were in general based on experimental evidence; however, it contained very little in terms of new knowledge[32].

By the 14th century, the use of the compass needle was widespread, and the phenomenon of magnetic attraction already established itself as a powerful image. This is exemplified by the allure of the voice of St. Giovanni Bonaventura, in the verse of the Canto XII, Paradise, in the Divine Comedy, written by the great Italian poet Dante Alighieri (1265–1321) in the first decades of the century*:

> "Out of the heart of one of the new lights
> There came a voice, that needle to the star
> Made me appear in turning thitherward."

* Translation by Henry Wadsworth Longfellow. In the Italian original:
"del cor de l'una de le luci nove si mosse voce,
che l'ago a la stella
parer mi fece in volgermi al suo dove"

In the 15th century, Nicholas of Cusa (Nicolaus Cusaneus) (1401–1464) in *Exercitationes*, likens Christ to the compass, both of them showing the way to men and both deriving their power from the heavens[33].

Wonderful is the Lodestone

In 1558 Giovanni Battista (or Giambattista) della Porta (1535–1615), a Neapolitan researcher and playwright, published the first edition of *Magia Naturalis* ('Natural Magic'), a work containing a mixture of natural facts, astrology, and practical knowledge. The second expanded version appeared in 1589, incorporating more scientific facts, including experiments with the lodestone. Della Porta was the first to propose a way of measuring the strength of the magnet, using a balance. The force exerted on a piece of iron was compensated with a weight put on another pan of the balance, and this weight was the measure of the lodestone's attraction. Della Porta was also a precursor in founding the first association for the study of scientific matters, the *Academia Secretorum Naturae* ('Academy of the Secrets of Nature'); their members called themselves *Otiosi* ('Leisure Men'). It was followed by the foundation in Rome of the Accademia dei Lincei, the first true scientific society, in 1603; Galileo Galilei was one of its members.

Near the end of the 16th century, Robert Norman (fl. 1590), English instrument maker, published in London *The New Attractive*, in 1581 (Figure 2.3); the book, on magnetism, was reprinted three times before the end of the century. Norman had been a sailor for twenty years and was well acquainted with the practical use of the compass. He wrote of the magnetic compass[34]:

> I guide the Pilot's course,
> his helping hand I am,
> The Mariner delights in me,
> so doth the Merchant man.
>
> The lodestone is the Stone,
> the only stone alone,
> Deserving praise above the rest,
> whose virtues are unknown.

He promises in the preface to "ground his arguments onlye upon experience, reasons and demonstrations"[35]. In this book he described

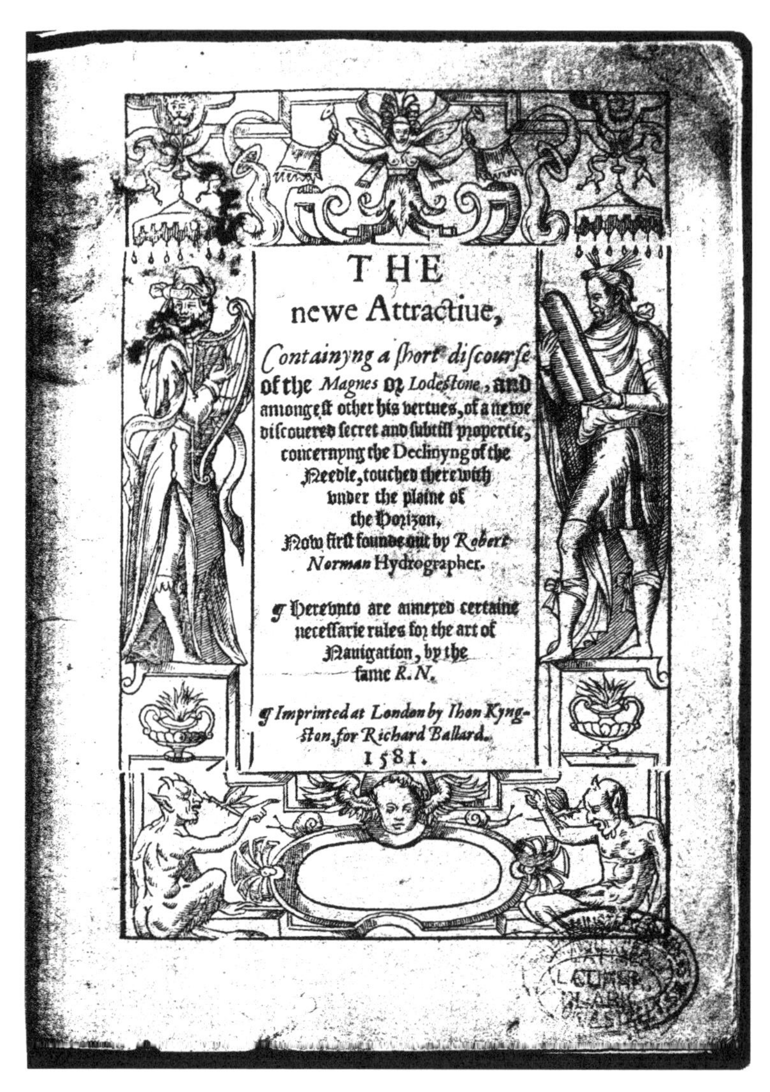
THE
newe Attractiue,
Containyng a short discourse
of the Magnes or Lodestone, and
amongest other his vertues, of a newe
discouered secret and subtill propertie,
concernyng the Declinyng of the
Needle, touched therewith
vnder the plaine of
the Horizon.
Now first founde out by Robert
Norman Hydrographer.

¶ Hereunto are annexed certaine
necessarie rules for the art of
Nauigation, by the
same R. N.

¶ Imprinted at London by Ihon Kyng-
ston, for Richard Ballard.
1581.

Figure 2.3 Title page of The New Attractive (1581), by the English mariner and instrument maker Robert Norman (fl. 1590) (courtesy Bakken Museum).

the magnetic dip, or inclination of the compass in relation to the horizontal, which he had discovered in 1576. In his own words, "(...) a newe discovered secrete and subtill propertie, concernying the declinying of the needle, touched therewith under the plaine of the hori-

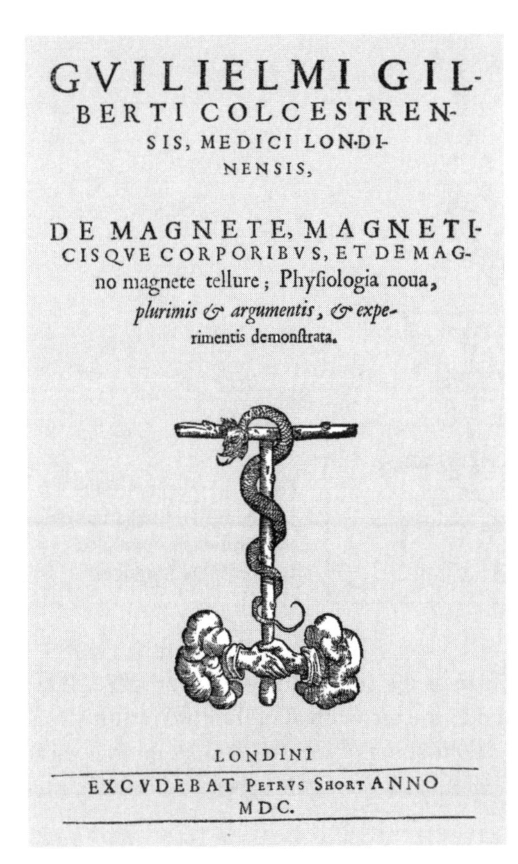

GVILIELMI GIL-
BERTI COLCESTREN-
SIS, MEDICI LONDI-
NENSIS,

DE MAGNETE, MAGNETI-
CISQVE CORPORIBVS, ET DE MAG-
no magnete tellure; Phyſiologia noua,
plurimis & argumentis, & expe-
rimentis demonſtrata.

LONDINI

EXCVDEBAT PETRVS SHORT ANNO
MDC.

Figure 2.4 Title page of De Magnete (On the magnet) (1600), by William Gilbert (1544–1603), the first treatise on magnetism (courtesy Bakken Museum).

zon."[36] This effect of inclination in relation to the horizontal plane had already been found before by the German instrument maker Georg Hartmann (1489–1564), in 1544.

Norman performed many experiments with compasses, and his discoveries afforded him "incredible delight". He thought that all compasses placed on different points of the Earth surface would point to the same "point respective". This conviction represents an important shift from Peregrinus' "power of the heavens", relating for the first time the influence upon the magnetic needle to the planet Earth.

The most important treatise on magnetism in this period, and for many years to come, was the book *De Magnete* ('On the Magnet')* (Figure 2.4), published in London in 1600 by William Gilbert (1544–

* For full title see Further reading.

Figure 2.5 William Gilbert (1544–1603), English physician and experimentalist, author of De Magnete (1600).

1603) (Figure 2.5), physician at the court of Queen Elizabeth I. Elizabeth I reigned in England in the second half of the 16th century, in a period characterized by a flourishing of the arts and literature, when England also reached a zenith in its economic power and influence in the world.

Elizabethan London witnessed, at about the same time that *De Magnete* was published, the first presentation of *Hamlet* (1600 or 1601), one of the greatest creations of the playwright and poet William Shakespeare (1564–1616). *Hamlet* tells the story of a Danish prince, based on a legendary figure, who is drawn into a series of tragedies as he tries to revenge the murder of his father. The words of one of Shakespeare's characters, John of Gaunt, Duke of Lancaster, uncle of Richard II, are appropriate to describe England in this period of glory,*

> "This royal throne of kings, this scepter'd isle
> This earth of majesty, this seat of Mars
> This other Eden, demi-paradise"

William Gilbert was born in 1544 in Colchester (Essex), northeast of London. He studied in Cambridge, where he obtained his BA in 1561

* W. Shakespeare, 'Richard II', Act 2, Scene 1.

and graduated in medicine in 1569. He was a talented experimentalist and dedicated himself to the study of electrical and magnetic phenomena; he created the word 'electric', from *elektron*, the Greek for amber. He was appointed president of the Royal College of Physicians and became the personal physician of Queen Elizabeth I in 1600. Besides *De Magnete* he wrote another book on cosmology, *De Mundo** (1651), published posthumously in Amsterdam by his half-brother, who had collected his scientific papers.

De Magnete is one of the major works in the history of the sciences because it represents an important milestone in the revolution of the attitude toward nature and the sciences, which occurred in the 16th and 17th centuries. Gilbert created a complete treatise of magnetism, but his aims were higher than that: he expected to inaugurate a new cosmology, in which magnetism played a central role, or a new philosophy of nature, or physiology, as he called it.

The work was written in Latin, and was divided into six books, each book into several chapters. In the Preface, he announces, to leave no doubts as to his allegiance to experimental science, that the work is dedicated to those "who not in books but in things themselves look for knowledge"[37]. In the first chapters of Book I he makes a critical revision of the writings on magnetism, condemning the myths that passed unchecked from one author to the next, tales that would not survive empirical tests, "figments and falsehoods" "dealt out to mankind to be swallowed"[38]. Among these "falsehoods", he lists the stories that account for ships built with wooden pegs to avoid the power of magnetic attraction of some mountains in the north, the destruction of magnetism by the contact with garlic, stories about the use of magnets to reveal unfaithful wives, and so on. He returns to these and other myths in Book II; he describes fables such as those mentioned by Pliny the Elder and by the Dominican bishop and philosopher Albertus Magnus (1200–1280), who believed in rocks that attracted flesh and wood, or others that report on rocks that had the power to attract gold[39]. He also reasoned against the idea of making perpetual motion machines with magnets, as had been suggested by Peregrinus and others.

He conducted many experiments with a spherically machined lodestone; since it represented a model of the Earth, he called it "terrella",

* Full title: *De Mundo Nostro Sublunari Philosophia Nova.*

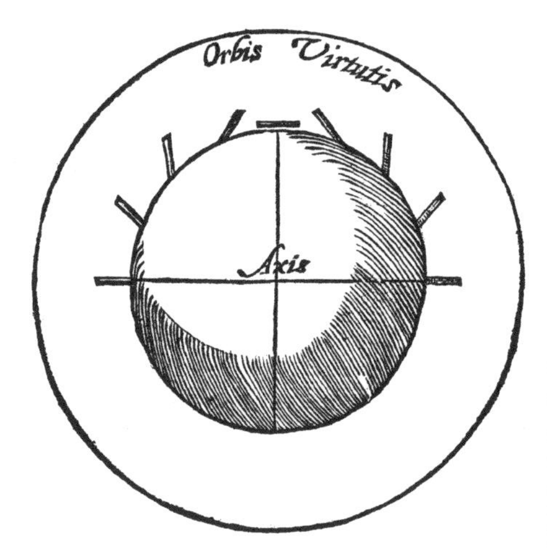

Figure 2.6 Drawing of a terrella with compasses on different points of its surface, showing the effect of inclination (or "dip"), from De Magnete (1600) (courtesy Bakken Museum).

or small Earth, an early example of an experimental scale model (Figure 2.6). He described how the poles of the *terrella* could be determined, following the same methods found in the text of Peregrinus. He experimented with the lodestone, showing that a split magnet produced two new magnets; he proved that the lodestone did not attract other metals than iron, and that it did not affect wood, glass and bone.

Gilbert compared the alignment of the compass in the North-South direction with the fact that the Earth's axis of rotation had a constant direction in space (neglecting other movements, such as precession and nutation): "Like the earth, the loadestone has the power of direction and of standing still at north and south; it has also a circular motion to the earth's position, whereby it adjusts itself to the earth's law."[40]

For him, there exists a link between the Earth's magnetism and the magnetism of the lodestone extracted from the mines: "Thus every separate fragment of the earth exhibits in indubitable experiments the whole impetus of magnetic matter; in its various movements it follows the terrestrial globe and the common principle of motion."[41]. And "Yet the loadstone and all the magnetic bodies – not only the stone but all magnetic, homogenic matter – seem to contain within themselves the potency of the earth's core and of its inmost viscera (...)."[42]

In Book II, he discusses several phenomena connected with magnetism, under the general denomination of 'movements'. These are: 1) Attraction, which he terms coition (from the Latin *co* + *ire*, going together); 2) The alignment with the North-South direction; 3) Declination, or deviation from the meridian, which he characterizes as a 'perverted' motion; 4) Magnetic dip, or inclination below the horizontal plane, and 5) Revolution, or circular motion[43].

In the same Book II, he describes electric phenomena, making *De Magnete* the first published treatise on electricity. In his study, he starts with the properties of amber; all the materials that attract chaff when rubbed he calls 'electrics'; the others were 'non-electrics'. According to Duane H. Roller, author of a study on *De Magnete*, with the establishment of this division Gilbert separated magnetic and electric phenomena, which existed side by side as effects related to 'attraction'. With this separation he founded the science of electricity[44].

In *De Magnete*, the electric attraction is demonstrated experimentally through its effect on a needle (not a magnetic one) mounted on a pivot, a simple instrument known as a *versorium*. Gilbert observed that this attraction of electric origin is reduced in cloudy weather, or when the object is exposed to the 'moisture from the mouth'. What he had thus proved was, putting it in modern terms, that moisture reduced the insulation of the object, allowing the electric charges to flow away, and therefore discharging it.

Gilbert tended to believe that the explanation for the electric attraction is due to some sort of material emission from the amber: "Hence it is probable that amber exhales something peculiar that attracts the bodies themselves, and not the air."[45] He proves that the intervening air is not involved, by demonstrating that the flame of a candle is not affected by the proximity with the attracting amber[46]. He criticizes earlier attempts by Epicurus, Plato, Galen and others, to explain magnetic attraction that used the same idea of 'effluvia'; for him, magnetic attraction (or rather, *coition*) results from the mutual action of lodestone and iron[47].

He also proves experimentally that heat momentarily destroys the magnetic properties of objects otherwise attracted by a magnet: a magnetized needle stands still near a piece of red-hot iron. He finds that the same piece of iron has its magnetic properties restored as it cools down.

He experiments with lodestones of different shapes and concludes that oblong ones exert a stronger attraction than the spherical ones. He divides a lodestone into two parts and finds that "If magnetic bodies be divided or in any way broken up, each several part hath a north end and a south end"[48]; re-joining two halves of a round lodestone cut along a parallel re-establishes the original magnet, with the poles in the original places.

Gilbert speaks of the property of the space around a lodestone: in modern terms, we would call it the magnetic field. He called this sphere of magnetic influence the *orbis virtutis*. He finds that inserting either boards, pottery, or marble, between the magnet and a piece of iron does not interfere with the attraction; only a plate of iron can affect the influence of the magnetic force. In his studies, he also notes that the magnetic effect of the lodestone is reduced as the distance is increased.

Gilbert discusses at length one of the most important properties of the magnetized bodies – their tendency to alignment in the North-South direction: "A little iron bar – that soul of the mariner's compass, that wonderful director in sea voyages, that finger of God, so to speak – points the way and has made known the whole circle of earth, unknown for so many ages."[49] And more: "(...) no invention of human arts has ever been of greater use to mankind"[50]. He challenges the authors that have claimed that the compass needle pointed to the celestial poles, or to the tail of the Bear in the constellation of the *Ursa Major.* He also contests the belief that by symmetry to what happens in the northern hemisphere, the compass in lands south of the equator will point the same end to the South Pole.

Gilbert's experiments demonstrated that a floating magnetic needle is not attracted as a whole to the Pole, repeating the experiments and conclusions of Robert Norman. In Gilbert's words[51] the "direction is not produced by the attraction, but by a disposing and conversory power existing in the earth as a whole (...)". In modern terms that fact might be explained as an effect arising from the action of two equal forces applied at different points (the poles of the magnetic needle), that turn the needle, but do not move the center of mass of the compass.

He considered several circumstances under which iron is naturally magnetized due to the influence of the Earth. For example, by allowing a piece of white-hot iron aligned in the North-South direction to cool; or by hammering or stretching an iron bar in this same direction. The same phenomenon is observed with a piece of iron that is part of the structure of a building, kept pointing along the North-South direction for many years.

In *De Magnete,* Gilbert studied the phenomenon of declination, or the deviation of the compass from the true North-South direction, by experimenting with the *terrella*. He proved that in the *terrella*, deviations occur when there are large irregularities on its surface; the same idea must then be applicable to the Earth, he argued. Again using the *terrella,* he demonstrates that the compass needle is parallel to the surface of the lodestone at a point near the magnetic equator, and is perpendicular to the surface at the poles. This deviation from the plane parallel to the surface (or horizontal direction) – the dip – is therefore beautifully explained. The angle formed by the needle with the horizontal, at intermediate latitudes, has a value between zero and ninety degrees (Figure 2.6).

In another experiment, he moved the magnetic needle away from the *terrella* and recorded the directions of the needle at several points. The figure representing the directions chosen by the needle located at circles of different radii is a rough anticipation of the images of the magnetic lines of field obtained by Michael Faraday using iron filings, more than 200 years later. In a remarkable comment on the abstract nature of this concept we would describe today as field lines, he states "(...) and so the spheres are magnetical, and yet are not real spheres existing by themselves"[52].

The rotation of the Earth is related in many ways to magnetism, according to Gilbert; he deals with this question at length in Book VI. He also discusses the application of magnetism to different questions, his "magnetic philosophy", or cosmology. "The causes of the diurnal motion are to be found in the magnetic energy and in the alliance of bodies"[53], he explains, "(...) produced partly by the energy of the magnetic property and partly by the superiority of the sun and his light"[54].

He rejects the Aristotelian belief that the Earth remained immobile in the center of the universe. He argues that the apparent turning of the stars is due to the rotation of the Earth: "it is more accordant to reason that the one small body, the earth, should make a daily revolution than that the whole universe should be hurled around it."[55] Without this rotation, one side of the earth would freeze with intense cold, and the other would be scorched, making life on the planet impossible[56]. He also recognizes the importance of the angle between the axis of the rotary motion of the Earth and the line perpendicular to the plane that contains the planet and the Sun; without this angle, there would be no seasons[57]. He is also aware that the axis of the Earth does not in fact point along a fixed direction, but turns in a slow motion called the precession of the equinoxes[58].

The names the poles of the magnetic compass used by Gilbert follow a convention different from that adopted nowadays. We now designate 'North' the end of the magnetic compass that points to the Earth's North Pole – Gilbert called it 'South'. It follows from the present nomenclature that the Earth's magnetic pole in the northern hemisphere is in fact a South magnetic pole.

In Book IV Gilbert discusses declination, which he attributes to irregularities in the surface of the Earth, not to magnetic mountains, as often held at the time[59]; he believed to have demonstrated this effect with a *terrella* that presented a depression on its surface[60]. The present explanation is more complex, and is related to the mechanisms acting inside the Earth's core that give rise to the magnetic properties of the planet (see Chapter 6).

Some of Gilbert's ideas found opposition in contemporary authors; for example, the Jesuit scholar Athanasius Kircher (1601–1680), an opponent of the Copernican heliocentric theory (enunciated in 1543), did not accept Gilbert's identification of the Earth with the magnet. He thought that if this was the case, the magnetic attraction would be so strong men would not be able to use iron tools; in the book *Magnes, sives De Arte Magnetica* ('The Magnet, or About the Magnetic Art') (1641) he qualified Gilbert's opinion as "*absurda, indigna et intolerabilis*"[61].

Gilbert presented a theory of the origin of iron, arising from a single element – 'earth'; the metallic properties of iron were produced by moist 'exhalations'. Although he used these Aristotelian terms, he gave them different meanings[62]. Furthermore, he considered iron, rather than silver or gold, the perfect metal, or the "foremost of metals". This is justified both on the grounds of his theory of metals, since iron is "earth in its own nature true and genuine", and also from the uncountable uses of the metal. Here one has an interesting example of a consideration of the social function of the metals entering the discussion of mineralogy or chemistry[63].

In his view of the universe, Gilbert rejects the existence of rigid spheres where stars and planets would be contained; it is implicit in his analysis, and evident in the diagram he presented in his later book *De Mundo* (1651)[64], that the universe is infinite. By extending the magnetic motive force to other celestial bodies besides the Earth, Gilbert transcends the boundaries between the sublunary and superlunary spheres, producing a "unified physics", with laws valid for the whole universe[65]. This is apparent, for example, when he abandons the idea of "heaviness", discarding this as something only related to the Earth, maintaining that "lunar bodies tend to the Moon, solar to the sun, within the respective spheres of their effluences"[66].

Gilbert's cosmology does not make an explicit choice in favor of a heliocentric solar system, since he speaks only of the movement of the Earth's rotation, not of the motion around the Sun. However, this choice is consistent with the scheme of heavens that will be later presented in *De Mundo*.

Gilbert's ideas influenced Johannes Kepler (1571–1630), who considered them one of the three fundamental elements for the formulation of his theory of the orbits of the planets, together with the heliocentric hypothesis of Copernicus and the astronomical data of Tycho Brahe (1546–1601)[67]. "If I believe anything," he wrote, "you after reading my book will be persuaded that I have placed a celestial rooftop upon the magnetical philosophy of Gilbert, who himself has built the terrestrial foundation"[68].

Gilbert's *De Magnete* represented an important step in the direction of creating a new paradigm for the sciences. His systematic use of the experimental method, his critical attitude toward the classics led him to innovate, stimulating many other authors. *De Magnete,* however, had a somewhat archaic language in some parts. For example, in places it showed an animistic inclination: "Wonderful is the loadstone shown in many experiments to be, and, as it were, animate", writes Gilbert[69]. There are also references expressing approval on comments by the ancient philosophers on the existence of a soul in the magnet, and in the world itself. He closes these references justifying the famous remark attributed to Thales on the soul of the magnet (see Chapter 1). Besides these passages on the souls, Gilbert often employs a vitalistic language as, for example, when he speaks of the increasing response of the iron to the attraction of the lodestone, as it gets nearer, describing it as "excited"[70]. *De Magnete* also lacks more elaborate quantitative statements, an element which would gain increasing relevance in the evolution of the new scientific method.

Only twelve years before the publication of *De Magnete,* the Great Spanish Armada had been defeated by the English Navy, an historical episode that changed the political map of Europe, and one to which the English cast-iron cannons made a significant contribution. At that time, only England dominated the technique of manufacture of such weapons, which substituted the bronze pieces. It is perhaps no coincidence that William Gilbert's book was published at such a time, when England's eminence was connected with the mastering of the nautical techniques and of iron manufacture[71]. It has been pointed out by the German science historian Edgar Zilsel that a large proportion of *De Magnete* deals with mining and metallurgy, 13% discusses nautical instruments, and 12% navigation in general[72].

After Gilbert's death in 1603, presumably of the plague[73], his books, stones and instruments were donated to the Royal College of Physicians; unfortunately, everything was lost when the College was destroyed in the Great London Fire of 1666. Much was lost also from the partial destruction of Colchester, Gilbert's birthplace, when it was under siege during the Second Civil War, in 1648[74].

For his great contribution to magnetism, William Gilbert was honored with the naming of the unit of magnetomotive force, magnetic analogue to the electromotive force; one gilbert (symbol Gi) measures the magnetomotive force corresponding to 0.75 amperes in the International System of Units (SI). In the words of the great English poet and dramatist John Dryden (1631–1700)[75]

> '*Gilbert* shall live, till *Load-stones* cease to draw,
> Or *British* Fleets the boundless ocean awe.'

Gilbert and the Scientific Revolution

The change in scientific attitude heralded by *De Magnete* merged with other influences and became part of a real scientific revolution, associated with the names of René Descartes, Francis Bacon and above all, Galileo Galilei*. This transformation of the scientific practice and

* A. C. Crombie summarizes the contribution of Galileo to the scientific method in the following words: "The special contribution that Galileo's conception of science as a mathematical description of relations enabled him to make to methodology, was to free it from a tendency of excessive empiricism which was the main defect of the Aristotelian tradition, and to give it a power of generality which was strictly related to experimental facts to a degree which previous Neoplatonists had seldom achieved."A. C. Crombie, *Robert Grosseteste and the Origins of Experimental Science 1100–1700*, Clarendon Press, Oxford, 1953, pg. 305.

world view would reach its climax in the second half of the 17th century with the English physicist Isaac Newton (1642–1727).

The early precondition for this revolution was the growing tendency to separate theology from science (*scientia*) towards the end of the Middle Ages. Up to that moment, theology was knowledge that originated directly from God, and therefore was of assured certainty. Natural knowledge, on the other hand, was fallible, and was supposed to be deducible from general principles.

In the early Middle Ages, theology was the queen of the sciences, in the words of St. Augustine (354–430). Augustine thought that should philosophers teach anything "contrary to our Scriptures, that is to Catholic faith, we may without any doubt believe it to be completely false, and we may by some means be able to show this"[76]. The gradual separation of the reign of theology and the domain of the sciences, and their demarcation, is associated especially with two medieval philosophers, the Scot, John Duns Scotus (c. 1266–1308) and the English philosopher and Franciscan priest, William of Ockham (c. 1284–1349)[77]; one may say that the intellectual revolution that gave birth to the scientific method had its roots in the works of medieval thinkers.

On the European continent, *De Magnete* influenced many authors, especially Johannes Kepler and Galileo Galilei. However, the importance of the work of William Gilbert was felt more strongly in England, where he was regarded as the founder of the experimental method, and also the introducer of Copernican theory[78]. In 1657, the architect and astronomer Christopher Wren (1632–1723), in his inaugural speech as Professor of Astronomy at Gresham College, in London, referred to him as "the father of the new Philosophy"[79].

The Polish astronomer Nicolaus Copernicus (1473–1543) had proposed in his book *De revolutionibus orbium coelestium* ('On the Revolutions of the Celestial Spheres'), published in 1543, that the Earth and the planets turned around the Sun in circular orbits. The fact that the stars did not change position in the sky as the Earth moved around the Sun was justified by Copernicus on the grounds that the stars were much farther away than hitherto admitted. The idea of a much larger universe, with the Earth removed from its privileged position at the center, had a great intellectual impact, starting the scientific revolution. After Copernicus published his heliocentric theory, the decision on its truth became possibly the chief scientific problem of the late 16th and early 17th centuries[80]. This question occupied the minds of

Figure 2.7 Galileo Galilei (1564–1642), Italian physicist and astronomer who was a pioneer in the mathematical formulation of the laws of physics.

the greatest thinkers of the period, such as Johannes Kepler, René Descartes and Galileo Galilei.

The scientific revolution gained momentum as men of learning started to study, with a fresh approach, the phenomenon of the movement of the bodies. The first step in this direction was to abandon the attitude of the Greek philosophers, who dealt with motion as a quality essential to the moving bodies, not as a state in which bodies happened to be. The question asked was not any longer how to define or characterize the essence of motion, but instead, how to describe motion in mathematical terms. This shift in emphasis occurred in parallel with advances in mathematics, more specifically in geometry, algebra and in mathematical notation; by the first decades of the 17th century, algebra and arithmetic had adopted essentially the same notation as employed today[81].

Galileo Galilei was born in Pisa, in 1564 (Figure 2.7), where he studied medicine; he later became lecturer at the university of Pisa, and in 1592, in Padua. He treated the problem of the motion of bodies, distinguishing between the conditions of uniform velocity and accelerated motion. Experimenting with balls rolling on planes, and observing that on smooth planes the motion continued with almost constant velocity, Galileo made the abstraction that in the absence of friction the velocity would remain constant forever. He concluded: "this particle will move along this same plane with a motion which is uniform and perpetual, provided the plane has no limits."[82] This was an early formulation of a principle that became known as the principle of inertia;

it represented an idea directly opposite to the Aristotelian conception that a body in motion required the constant application of forces.

Galileo Galilei intended to treat the motion of bodies mathematically, emulating the work Archimedes (287–212 BC) had carried out in antiquity with the study of forces, with applications to the lever and to the balance. Galileo's studies of kinematics benefited from his emphasis on measurement and mathematical analysis of the physical facts; the necessary mathematical tools were already available at the time. His results were reported in the dialogue and treatise of the same title – *De Motu,* ('About Movement'), written about 1590.

Galileo was fascinated with the magnetism of the lodestone, and conducted many experiments to investigate its properties. Galileo believed that forces similar to magnetic forces acted in space, between the heavenly bodies, and that the Earth was a huge magnet[83]. He read *De Magnete* and was much impressed with its content and methodology; he mentions it in his most complete work, the *Discorsi e dimostrazioni mathematiche intorno a due nuove scienze attenenti alla meccanica* ('Dialogue Concerning Two New Sciences'), published in 1634. In the words of Salviati, who represents Galileo in the dialogues, he would have missed Gilbert's book, "if a famous Peripatetic [Aristotelian] philosopher had not made a present of it, I think in order to protect his library from its contagion"[84]. The other character of the Dialogues, Simplicius, asks Salviati: "Then you are one of those people who adhere to the magnetic philosophy of William Gilbert?" and Salviati replies: "Certainly I am, and I believe that I have for company every man who has attentively read his book and carried out his experiments."[85]

However, Galileo also wrote[86]: "I want to tell you about one particular to which I wish Gilbert had not lent his ear", referring with disapproval to the suggestion that Gilbert had made (taken from Peregrinus) that the *terrella* would turn by itself, if aligned with the axis of the Earth.

In a letter to William Barlowe (d. 1625), written between 1600 and 1603, Gilbert acknowledges correspondence from Joannes Franciscus Sagredus (1571–1620) which he had received through the Venetian ambassador: "Sagredo is a great Magnetical man and writeth that he has conferred with ... the Readers of Padua and reported wonderful liking of my booke". Sagredo was a friend of Galileo, and Galileo was a reader at Padua, therefore he certainly was among those that displayed such "wonderful liking"[87].

Having heard of the invention of the telescope in Holland, Galileo built one himself in 1609, calling it *perspicillum*; the first version magnified only three times, but as Galileo refined his technique, he produced much better instruments. One may say that he thus built some of the first scientific instruments; his observations with the telescope would revolutionize man's understanding of the universe. One can imagine the emotion of Galileo as he pointed his telescope to the night sky for the first time. He was stunned to perceive on the surface of the Moon, previously regarded as smooth, mountains that resembled those on Earth; near Jupiter, satellites that turned around the planet; spots on the face of the Sun; thousands of stars never previously imagined, and the image of Venus showing phases like the Moon! About the Moon he wrote[88]: "It is a most beautiful and delightful sight to behold the body of the moon ..."

His fantastic discoveries were communicated to the world in his book *Sidereus Nuncius* ('The Starry Messenger'), published in 1610. The impact of this publication on his contemporaries was very great, enhanced by the fact that Galileo chose to write his books in Italian, instead of Latin.

The presence of mountains on the Moon and spots on the Sun's surface helped to destroy the idea of supralunary and perfect spheres, where the stars and the Sun were located, and which differed from the Earth, the realm of corruption and imperfection. This inaugurated a picture of a unified world, and meant the end of the idea of Cosmos, or a finite, closed, hierarchy of spheres. The importance of this transformation is characterized by the science historian Alexandre Koyré in strong words: "The dissolution of the Cosmos – I repeat what I have already said: this seems to me to be the most profound revolution achieved or suffered by the human mind since the invention of the Cosmos by the Greeks."[89] This revolution would have its martyrs, like the philosopher Giordano Bruno (b. 1548) who was condemned by the Inquisition for his philosophical ideas, his heliocentric views, and his belief in an infinite universe, and burned at the stake in Rome in 1600.

Galileo too was persecuted by the Church for his teachings on the Copernican worldview, and was condemned by the Inquisition, and sentenced to spend his last years of life under house arrest. During this period, he was still intellectually very active, and published his last book, 'Dialogue Concerning Two New Sciences'. He was forced to recant his beliefs, under the threat of torture: he concluded his statements with the words: "I, Galileo Galilei, have abjured as above with my own hand."[90] Galileo was 'rehabilitated' by the Church only 360 years later, in 1992.

Galileo Galilei died in 1642, the same year that Isaac Newton was born. Isaac Newton would give the definitive form to the laws of motion proposed by Galileo. He would complete the revolution in world picture, unifying the laws of physics that acted on the solar system and on the Earth (see Chapter 4).

René Descartes, another of the founders of the New Science, was born in La Haye (now Descartes), in France, in 1596. He studied law in Poitiers and later lived for many years in the Netherlands. Descartes was one of the chief advocates of the new picture of the universe, where the same laws ruled terrestrial and celestial phenomena. In his book *Le Monde* ('The World), written in the early 1630s, he attempts a mechanical explanation of the universe, avoiding scholastic concepts. Although Descartes made very important contributions to mathematics and to mathematical techniques used in physics, he expressed his ideas in physics mostly in non-mathematical terms, and insisted on the search for intrinsic or essential causes; he remained a philosopher dealing with physical problems. In that approach, he differed from the way Galileo treated the issues of the physical world. For example, commenting on the work of Galileo, he wrote: "As to what Galileo has written about the balance and the lever, he explains very well what happens (*quod ita fit*), but not why it happens (*cur ita fit*), as I have done in my *Principles*[91]".

The use of mathematics in the treatment of the matters of physical reality also separated Galileo from the author of *De Magnete*. Speaking of William Gilbert, Galileo wrote: "I have the highest praise, admiration, and envy for this author, who framed such a stupendous concept regarding an object which innumerable men of splendid intellect had handled without paying any attention to it (...) What I might have wished for in Gilbert would be a little more of the mathematician, and especially a thorough grounding in geometry, a discipline which would have rendered him less rash about accepting as rigorous proofs those reasons which he puts forward as *verae causae* [true causes] for the correct conclusions he himself had observed."[92]

The mathematization of the world view, implicit in the method of Galileo, has its roots in the teachings of the Pythagoreans. Plato, who lived from 427 to 347 BC, continued this tendency; for him, the structure of the universe was related to mathematical entities in an intimate way: particles of fire were tetrahedra; air was formed of octahedra, and so on. The universe itself had spherical shape, since the sphere is the most perfect of forms. The point of view of Plato and followers contrasts with the vision of Aristotle in many fundamental ways, and also in this specific point; Aristotle framed his arguments in qualitative, not quantitative terms.

Adherence to this view usually leads to the choice of mathematics as the language most adequate to express the facts of nature. Galileo Galilei, at the beginning of the 17th century, conveys[93] this view in his book *Sidereus Nuncius*: "Philosophy [nature] is written in that great book which ever lies before our eyes. I mean the universe, but we cannot understand it if we do not first learn the language and grasp the symbols in which it is written. The book is written in the mathematical language, and the symbols are triangles, circles and other geometrical figures without whose help it is humanly impossible to comprehend a single word of it, and without which one wanders in vain through a dark labyrinth." This concept reached maturity only with the development of other mathematical techniques used to describe more complex physical processes, such as the differential calculus, invented by Newton and Leibniz, in the 17th century.

The astronomer Johannes Kepler, born in 1571 in Weil-der-Stadt, Swabia, Southern Germany, solved the problem of the orbits of the planets. Holding the Copernican view that the Earth and the other planets turned around the Sun, and using the astronomical observations of the Danish astronomer Tycho Brahe (1546–1601), Kepler discovered that the planets described elliptical, not circular orbits. Inspired by the "natural philosophy" of Gilbert, he considered that the planets were maintained in their motion by magnetic forces.

The scientific revolution would complete its cycle with the mechanics and the theory of gravitation of Isaac Newton. The laws of Kepler were then shown to fit into a scheme of forces of gravitational attraction that fell with the square of the distance. These forces determined the orbits of the planets, satellites and comets.

Summing up the accomplishments of science in the 17th century, the American astronomer Herbert Dingle (1890–1978)[94] wrote: "The achievement of the seventeenth century had been precise profound and immeasurably great. At the beginning of that great epoch the conception of the universe which, though not unchallenged, was still predominant, was essentially medieval in spirit; at the centre were the earth and the baser elements, subject to change and decay and tending to move in straight lines towards their own places, while surrounding them the crystalline spheres, incorruptible in the heavens, performed eternally the perfect circular motions proper to their nature. By the end of the century the whole scheme had vanished and the Newtonian system reigned in its stead. The spheres had disappeared; matter and motion were the same everywhere, on the earth and in the heavens, and were indissolubly linked with one another through the universal force of gravitation; and a technique had been created for the conquest of other phenomena by means of Newtonian forces yet to be discovered."

Further Reading

Amir D. Aczel, *The Riddle of the Compass*, Harcourt, New York, 2001.

A.C. Crombie, *The History of Science from Augustine to Galileo*, Dover Publications, New York, 1995.

W. Gilbert, De Magnete, *Great Books*, Vol. 28, Encyclopaedia Britannica, Chicago 1978. The complete title is *De Magnete Magneticisque Corporibus et de Magno Magnete Tellure Physiologia Nova* ("On the Magnet and Magnetic Bodies, and on The New Physiology of That Great Magnet the Earth").

Pierre de Marincourt, "The Letter of Peregrinus", in *Source Book in Medieval Science*, ed. E. Grant, Harvard University Press, Cambridge, 1974.

Stephen Pumfrey, *Latitude & the Magnetic Earth*, Icon Books, Cambridge, 2002.

Duane H.D. Roller, *The De Magnete of William Gilbert*, Menno Hertzberger, Amsterdam, 1959.

Chapter 3

The Unification: Electricity and Magnetism

"From time immemorial, as long as there has been any natural science, its ultimate supreme goal has been the combination of the motley diversity of physical phenomena into a unified system or even a single formula; (...)"[1]

Max Planck, *in 'The Unity of the Physical World-Picture'.*

Electric Shock

Since antiquity, two instances of action at a distance have been known: the attraction of dust and chaff by a piece of amber rubbed by cloth or fur, and the more remarkable magnetic attraction of iron by the lodestone. In his Natural History, written in the 1st century AD, Pliny the Elder acknowledges in his notes the similarity between the two effects, since amber[2], "attracts straw, dry leaves and bark from the linden-tree, just as a magnet attracts iron".

The first phenomenon has an electrical origin; amber and the cloth, originally uncharged, acquire with rubbing charges of opposite signs: one develops an excess of electrons (particles of negative charge) and the other an electron deficiency (therefore, a positive charge).

Amber was well known in Greece from the 9th century BC[3], and from *elektron*, the Greek word for amber, derives the word electricity. A charged piece of amber induces an electric charge of the opposite sign on the nearest surface of the originally neutral chaff, and therefore the chaff is attracted by the amber, since opposite charges are attracted, and like charges are repelled. The presence of an electrically charged object can be detected using this property of repulsion of equal charges: the two halves of a thin folded leaf of paper or gold separate when in contact with a charged object. An instrument made with such a leaf enclosed in a glass jar is known as an electroscope, and has

From Lodestone to Supermagnets. Alberto P. Guimarães

ISBN: 3-527-40557-7

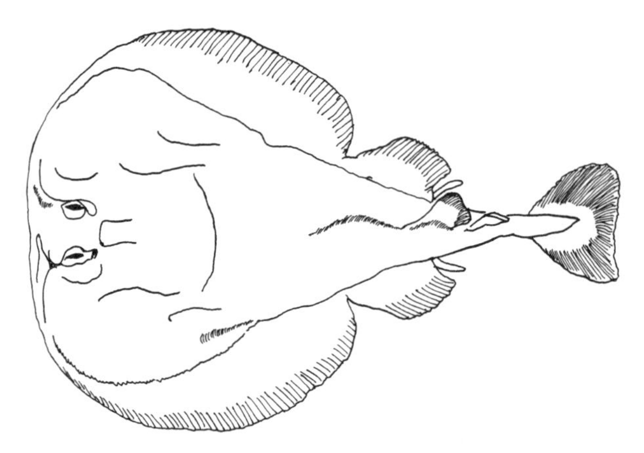

Figure 3.1 The electric ray Torpedo nobiliana, used in Greece and Rome for the treatment of several diseases.

evolved from a pair of linen threads first used for this purpose by Benjamin Franklin (1706–1790) in the 18th century.

The process of charging objects by friction relies on the possibility of displacing the elementary negative particles, which were identified in the 19th century with the electrons. This displacement is also at the basis of the charge transport in electrical conductors, like the copper wires of the electricity network. Until the 18th century, it was not evident that the effects of electrification by friction, the so-called static electricity effects, and the transport of current through conductors, were both due to the same charge carriers.

Another phenomenon of electric origin known in Greek and Roman antiquity is the power of some fish to induce numbness (*torpere*, in Latin), among these the rays called torpedoes (of the family Torpedinidae), of which several species occur in the Mediterranean (Figure 3.1). Plato refers to the electric ray in the dialogue where Meno tells Socrates that his "power over others to be very like the flat torpedo fish, who torpifies those who come near him and touch him"[4]. Aristotle, in his *History of Animals*, also described that the torpedo "narcotizes the creatures that it wants to catch, overpowering them by the power of shock that is resident in its body, and feeds upon them"[5]. The therapeutic use of these fish to cure headaches was recommended by Scribonius Largus (1st century AD), a Roman physician, who published about the year AD 47 the *Compositiones medicamentorum*[6]. The electrical character of the action of the electric rays was not discovered until the second half of the 18th century, after some studies of the South American electric eel *Electrophorus electricus*[7].

The first systematic study of the electrical properties of materials was published by William Gilbert, in his book *De Magnete* (1600), a treatise mostly devoted to magnetism (see Chapter 2). He made an important classification of substances according to their electric behavior; he determined which materials showed the same effect of electric attraction as amber, and called them *electrics*. In this class of substances he included glass, crystal, diamond, sapphire, sealing wax, and sulfur; the other class of materials, the *non-electrics*, for Gilbert, included the metals, for example.

Another early researcher of electricity was the Italian Jesuit Niccolo Cabeo (1586–1650), who published the book *Philosophia Magnetica* in 1629. In this work the fact that bodies with electric charges of the same polarity show a force of repulsion was enunciated for the first time: "filings attracted by excited amber sometimes recoiled at a distance of several inches after making contact"[8]. This phenomenon is due to the fact that the filings once in contact with the amber acquire a charge of the same sign, therefore the force of repulsion. This effect was discussed by the German physicist Otto von Guericke (1602–1686) in Book IV of his *Experimenta Nova (ut vocantur) Magdeburgica de Vacuo Spatio* ('New Magdeburg Experiments on Void Space'), published[9] in 1672. The English physicist Isaac Newton (1642–1727) also observed, around 1675, electric repulsion when he rubbed a piece of glass, close to some pieces of paper[10].

The British chemist Stephen Gray (1666?–1736) discovered in 1729 that by rubbing an object he could make another distant object attract light bodies, provided they were connected through some adequate materials; this effect proved the existence of electric conduction. Gray also discovered the induction of charges: a piece of lead hanging from a thread was electrically charged by approaching, without physical contact, a charged glass tube. These experiments were repeated in Paris by Charles-François Du Fay (1638–1739), who published his researches in 1733. He is the first person to correlate the two categories of materials created by Gilbert – *electrics* and *non-electrics* – with their degree of electrical conductivity[11]: the *electrics* were the materials that did not conduct electricity – the insulators, and the *non-electrics* were the conductors, or in his words, materials that "transmitted ... the electric matter“[12]. The denomination of “conductors” would be used later by Jean Théophile Desaguliers[13] (1638–1744). The non-conductors were called by Desaguliers “electrics per se“, or “supporters”.

From the analysis of attraction or repulsion between different electrified materials, Du Fay concluded that there exist two different types of electricity: “resinous electricity” and “vitreous electricity”. In modern terms, this meant that there are two classes of materials: some that, after being rubbed, have an excess negative charge (resinous), others an excess positive charge (vitreous). From this distinction, the picture of electric conductivity in terms of flow of *two* electric fluids resulted naturally, an idea that was for a long period adopted by the investigators, and is now known to be incorrect. Du Fay reported that he had observed vitreous electricity in glass, rock crystal, precious stones, wool; and resinous electricity in amber, silk, paper. However, other observers later found that depending on which material the vitreous or the resinous materials were rubbed with, they could show either type of electricity.

In the following years of the 18th century, it became clear that only one electric species (known today to be the negatively charged electrons), is transported in the phenomenon of electric conduction. This one-fluid hypothesis was first formulated by the American scholar and political leader Benjamin Franklin (1706–1790), who became famous for his experiments with atmospheric electricity made around 1750. Franklin found that the electric effects were due to the excess or deficiency of the electric fluid, also known at the time as “electric matter“, or “electric fire”. He demonstrated this fact[14] by charging two men standing on glass (insulating) stools; one was charged plus and the other minus. When they touched hands, there was a spark, since the charges flowed from one to the other, re-establishing neutrality in both. If either instead touched a third, uncharged, man, they would feel electric shocks, since electricity would flow to the one that had a deficiency of charge, or from the one that had an excess of it.

Franklin also explained that electric charge, or “electric matter“, was not *created* by rubbing objects, it was simply displaced from one object to the other. The charging of objects by rubbing is observed when the electric charges leave the rubbed object (glass, sealing wax, or any other electrically insulating material) to the cloth, or vice-versa. This and other related phenomena of electrical origin could not be studied in more systematic fashion until efficient forms of charging objects, and ways of storing the electric charge, had been devised. Until this happened, observers were practically limited to natural sources of electricity. Among these sources in nature, the phenomena related to the accumulation and flow of atmospheric charge stood out.

Lightning has been known for its frightening effects since the early days of the prehistory of humankind. This phenomenon arises from the accumulation of charges in the clouds, which may lead to huge differences of potential, of the order of a hundred million volts, either between different parts of the clouds, or between the clouds and the ground. Lightning is the very intense electrical discharge observed during thunderstorms that tends to equalize these potentials, with currents reaching typically ten thousand amperes, and lasting less than one tenth of a second. This is a fairly common phenomenon; at each given instant, some two thousand thunderstorms occur at different points of the Earth.

Although everyone is accustomed nowadays to attribute these natural phenomena to atmospheric electricity, before the beginning of the 18th century the fact that thunder and lightning had electrical origin was not known. In 1708, the English investigator William Wall produced sparks by sliding a long piece of amber through a woollen material and compared the appearance of the sparks, and the associated sounds, to those of atmospheric phenomena[15]. Benjamin Franklin was the first definitely to establish this connection. In his famous experiment with atmospheric electricity (1752), Franklin managed to use the line of a kite to conduct the electric charges from the atmosphere, which were then safely led to the ground; his studies resulted in the invention of the lightning rod.

The historical records are full of events involving atmospheric electricity, including many tragedies produced by lightning. Some observers attempted to study this force of nature more closely, and paid with their lives for their curiosity. One example is that of the Estonian physicist Georg Wilhelm Richmann (1711–1753), killed in St. Petersburg in 1753 by lightning, while studying atmospheric electricity. He had set up a conductor on the roof of his house, and was trying to measure the quantity of atmospheric electricity during a thunderstorm with an electroscope [16].

The need for sources of electric charges for the accomplishment of electricity experiments led many investigators to build machines that used friction to charge different objects. Many electrostatic machines were built; one of the first of such machines was invented by the German physicist Otto von Guericke (1602–1686) in 1663. Von Guericke is best known for the invention of the air pump, and for the experiments with evacuated vessels. He once evacuated two metallic hemispheres (the Magdeburg hemispheres) and showed that atmospheric pressure held them with such a force that they could not be separated by a team of sixteen horses.

Von Guericke's electrostatic machine consisted of a globe of sulfur on an iron axis around which it could turn. Rubbing the turning globe with his hands, or with a piece of cloth, von Guericke charged the globe, and managed to make several experiments with electricity. He observed that the charged globe could induce a charge on other objects, and also that electricity could be carried by some bodies. The conduction of electricity was demonstrated by showing that electric effects, such as the attraction of small objects, would seem as if distant objects were physically connected to the charged globe.

In 1675 Jean Picard (1620–1682), a French astronomer at the Collège de France, the first to measure accurately the length of a degree of a meridian, made an interesting observation. He noticed, as he transported a barometer consisting of a glass tube filled with mercury during the night, the appearance of a blue light inside the tube[17]. This phenomenon was investigated systematically thirty years later by the English self-taught scientist Francis Hauksbee, the Elder (c. 1666-c. 1713); Hauksbee found then that the luminosity was due to static electricity produced by the friction of the mercury against the glass tube. Inspired by this discovery, he made an electrostatic generator consisting of a turning glass tube that was rubbed by his hands; he had re-invented von Guericke's electrostatic machine, using glass instead of sulfur.

Another important advance in the study of electrical phenomena was the accidental discovery of a suitable means of storing electrical charge, the Leyden jar, in Holland in 1746, by the Dutch physicist Pieter van Musschenbroek (1692–1761). This device was simultaneously invented by the administrator and cleric Ewald Georg von Kleist (c. 1700–1748), from Pomerania, in Prussia. The Leyden jar consisted initially of a glass jar filled with water, with a conducting cable immersed in it. This original design was later improved into a glass bottle lined with metal foil, both internally and externally. It works by storing opposite charges in the two metallic foils. The Leyden jar is a direct ancestor of the modern capacitor, a common component in any electronic circuit.

Many experiments in electricity were made by combining electrostatic machines with Leyden jars. In 1847, William Watson (1715–1787) and other fellows of the Royal Society managed to send an electric signal across the Thames, discharging a Leyden jar through the water of the river, by inserting two wires, one on each bank, 400 yards apart[18].

Leyden jars were used to produce electrical shocks in human subjects; this unpleasant sensation arises from the flow of electricity through the body tissues. Depending on the intensity of the current, and its path, electrical shocks may of course be lethal. Musschenbroek himself, reporting his discovery to the French physicist René-Antoine de Réamur (1683–1757) in 1746, wrote[19]:

> "I wish to report to you a new but terrible experiment, which I advise you on no account to attempt yourself."

In France, the electrical capacity of the Leyden jars was demonstrated in many ways; on one occasion, King Louis XV witnessed the discharge of a jar through a human chain of 180 Royal Guards holding hands. Many amateurs exhibited the remarkable effects of electricity in the eighteenth-century European salons.

The Electric Battery

A more effective form of inducing electrical phenomena had to wait until electricity was produced by chemical reactions. The first indication that electricity was made to flow as a consequence of these reactions was the series of experiments conducted by the Italian anatomist Luigi Galvani (1737–1798) (Figure 3.2) in the 1780s in Bologna.

The physiological effects of electric shocks, including the acceleration of the pulse, perspiration and contraction of the muscles led to the picture of electricity as a "vital force". The resuscitation or "reanimation" of drowned or apparently dead people was proposed in 1778 by Charles Kite at the Royal Human Society in London; those that did not respond to these stimuli were considered dead[20].

Galvani started a study of the response of nerves and muscles of frogs to static electricity in 1780. He connected an electrical machine

Figure 3.2 The Italian anatomist Luigi Galvani (1737–1798).

to the assembly of spinal cord, crural nerve and lower limbs of the frogs, and observed that the legs contracted when subjected to electric discharges. Then he found out that the legs contracted even when insulated from the machine, whenever the machine produced a spark. He tried the same experiment using atmospheric electricity, and again obtained a contraction.

An unexpected effect was observed when he hung the legs with brass hooks from an iron railing in the garden of his house. Galvani had observed that the legs contracted during a thunderstorm, but this time he found that they still twisted when the storm was over, when "the sky was quiet and serene"[21]. They showed the same behavior when taken indoors; the response was observed every time the leg was in contact with two dissimilar metals. As he touched the limbs with different metallic objects, Galvani noticed that the intensity of the response depended on which metals were employed.

Galvani (incorrectly) interpreted these observations as proof of the existence of some "animal electricity", *un'elettricità particolare*; he thought that the muscles of the dead animal stored an electric fluid produced by the brains that was responsible for these effects. When his observations were published in 1791 in the paper *De viribus electricitatis in motu musculari commentarius* ('Commentary on the Effect of Electricity on Muscular Motion') they produced a strong impact. Galvani was then professor of obstetric arts at the Istituto delle Scienze at the University of Bologna.

Figure 3.3 The Italian physicist Alessandro Volta (1745–1827), inventor of the electric battery.

Among those who reacted to these discoveries was the Italian researcher Alessandro Volta (1745–1827) (Figure 3.3); he did not have a high opinion of the physicians, whom he considered "ignorant of the known laws of electricity"[22]. Volta correctly explained the results of Galvani, pointing that they were due to the chemical reaction between the two metals through the moisture covering the frog legs. In modern language, one might say that when two metals are separated by a conducting liquid (called an electrolyte), a potential difference appears between them, an electric current flows, and this stimulates the nerves, and then the muscles in the frog limbs. The electrical potential had nothing to do with "animal electricity"; only its detection had to do with the sensitivity of the nerves of the frog to electrical stimulation.

The identification of electricity as the 'vital force', or its connection to the essence of life, remained in the imagination of authors throughout the 18th and 19th centuries. An expression of the forcefulness of this link is given by the novel *Frankenstein, or the Modern Prometheus* (1818), by the English novelist Mary Godwin Shelley (1797–1851). The suggestion made by the author is that "Perhaps a corpse would be reanimated; galvanism had given token of such things; perhaps the component parts of a creature might be manufactured, brought together, and endued with vital warmth“[23]. *Frankenstein* describes how the experiment of creating a humanoid from parts of corpses takes a terrible turn, after life is instilled into the monster by the application of a powerful electric discharge.

In our age, the link binding electricity to life is present in the technique of electric defibrillation: electric shocks are used to save the lives of patients whose hearts engage in an erratic electrical activity called fibrillation. The electric activity of living tissue is also at the basis of several important diagnostic procedures, for example, electromyography (EMG) studies muscle activity, electroencephalography (EEG) activity of the brain, and electroretinograms (ERM), the electric potentials associated with the stimulation of the retina.

Alessandro Giuseppe Antonio Anastasio Volta was a physicist, born in Como, Italy, in 1745. When he was thirty he invented the electrophorus, a static electricity generator formed of a disc made of resin and wax that, when rubbed, develops an electric charge; a metallic disc closely mounted acquires an induced charge. Volta conducted many experiments in electricity, and his most important contribution came when he was a Professor of Experimental Physics in Pavia, in the 1790s. He was a sociable man, and had other interests, judging from the comment of his friend Lichtenberg, according to whom Volta “understood a lot about the electricity of women“[24].

In the beginning of his search after the announcement of the discovery of Galvani, Volta took up the experiments of the anatomist “with little hope of success“[25], but in April 1792 he was able to repeat the same results, giving them the correct interpretation.

In the experiments with the frogs, the potential difference was detected by the nerves; Volta invented another method of detection. He found that two discs of dissimilar metals in contact with his tongue gave a strange feeling; the saliva was the intervening medium in this experiment. He did not know that this effect had been reported in 1762 by the Swiss Professor of Mathematics Johann Georg Sulzer (1720–1779), who, however, had not related it to electricity[26].

We now know that when two pieces of metal are separated by an electrolyte there is a chemical reaction that gives rise to a small potential difference, e.g. in the case of zinc and silver, of 0.78 volts. To multiply this potential difference, Volta piled several pairs of discs of these two metals, each pair separated by a piece of cloth or cardboard moistened in brine. This arrangement became known as the voltaic pile and constituted the first electric battery. When the two outermost discs were connected through a conducting wire, a flow of electric current was observed. This discovery was communicated in a letter in 1800 to Sir Joseph Banks (1743–1820), president of the Royal Society of London; it was published in the Philosophical Transactions of the Society in the same year.

Alessandro Volta demonstrated his discoveries at the Academy of Sciences of Paris, and they made a strong impression on Napoleon, who decided to award a gold medal "for the best experiment made each year on the galvanic fluid [electricity]", and a prize of 60 000 francs for contributions comparable to "Franklin's and Volta's"[27].

Volta's invention created the means of generating steady electric currents, which opened up a new world of experimental possibilities with electricity, explored in subsequent years, especially with the work of Oersted, Ampère and Faraday.

Volta's name was perpetuated in the name of the unit of potential difference, or tension, in the International System, the volt (abbreviated V).

It was finally established during the 18th century that frictional electricity, or static electricity, was the same as 'galvanic' electricity, or the electricity produced by Volta's pile. To understand why these phenomena sometimes presented a different form, one has to examine more carefully the meaning of the terms 'voltage' or 'potential difference', and 'current'. To understand these concepts, relevant to the study of electricity, a parallel is usually made between the electric circuit and a system of water pipes. The voltage or potential difference in electricity is analogous to the potential energy of a water reservoir placed at a certain height; the intensity of the electric current (or amount of charge circulating per unit time) is analogous to the water output (volume of water per unit time) flowing down from the reservoir through the pipe. In frictional or static electricity, the experiments usually involved high potential differences and small currents; sparks are only observed when the potential differences are high, of the order of hundreds or thousands of volts. The currents last only a short time, and the total charges transported in these experiments are usually small. The opposite is true of the effects related to the voltaic pile, which provided lower voltages and steady currents.

In an electric circuit, the intensity of the current is related to the voltage. For a given circuit, characterized by a certain resistance, the current is directly proportional to the voltage across it, and inversely proportional to the value of this resistance. This relationship, valid for most materials, is known as Ohm's Law, after the German physicist Georg Simon Ohm (1789–1854), a schoolteacher in Cologne who published it in 1827.

The Unity of Nature

Nature presents to the eye of the inquisitive observer a wide variety of processes and objects of study: the light-emitting Sun, the diversity of landscapes on the Earth, the Moon and the stars that decorate the night sky, the wandering planets, and so on. Several centuries before our era, pre-socratic philosophers in Greece looked for a unity in nature underlying this apparent wide diversity of natural phenomena. The existence of this unity, according to the early Greek thinkers, was intertwined with the concept of the universe as an organism[28]. This idea of unity is particularly clear in the teachings of the Pythagoreans*. Instead of unity, one can also speak, in line with these thinkers, of a kinship of nature[29].

Unity in this context is the unity in nature, not in science; in other words, the unity of the object of study of the sciences, something different from the unity in the corpus of science, posited by some philosophers of science. One may argue, however, that the unity of nature would imply unity in the sciences. In the 18th century, Roger Boscovich (1711–1787) (or Rudjer Josip Boscovic), a Croatian Jesuit born in Ragusa, present-day Dubrovnik, published (in 1758) his *Theoria Philosophiae Naturalis* ('Theory of Natural Philosophy') in Vienna, and his project was to "derive all observed physical phenomena from a single law"[30]. This is the same aim that modern physicists pursue under the over-ambitious name of 'Theories of Everything'. One of these theories is the string theory of the particle physicists, which attempts to describe in a unified way all the physical interactions (or

* According to Guthrie: 'The world is divine, it is therefore good, and is a single whole. If it is good, alive and a whole, that is because, said Pythagoras, it is *limited*, and displays an *order* in the relations of its various parts. Full and efficient life depends on organization.' W.K.C. Guthrie, *The Greek Philosophers – from Thales to Aristotle*, Methuen & Co, (London 1967), p. 37.

'forces'), including gravitation and electromagnetism[31]. These strings are elementary entities that can vibrate, and one can in a way think that the 'notes' of these vibrations are the particles, such as the electrons. However, there are many problems with this theory, one of them being that it requires a universe of ten dimensions, with seven of these practically inaccessible to experiment.

In an attenuated form, the goal of attaining a unified picture of the physical world is shared by all physicists. The German physicist Carl Friedrich von Weizsäcker, known as a proponent of a theory of formation of the solar planets, considers "the endeavour of physics to achieve a unified world view. We do not accept appearances in their many-coloured fullness, but we want to explain them, that is, we want to reduce one fact to another"[32]. A similar view, given in the opening quotation of this Chapter, was expressed in the 20th century by the German physicist Max Planck (1858–1947), the founder of quantum mechanics, the physics that describes the atomic and subatomic world.

German *Naturphilosophie* (Natural Philosophy), a tendency among some philosophers, rather than a school of thought, flourished in the early 19th century, and is associated with the name of the philosopher Friedrich Wilhelm Joseph von Schelling, born in 1775 in Leonberg, Germany. He wrote *Ideas on a Philosophy of Nature* (1797), where he exposed his conception of the subject.

Schelling was influenced by the ideas of the philosopher Immanuel Kant (1724–1804), and posited a basic unity in nature. Kant, probably the greatest philosopher of modern times, had written that "(...) all natural philosophy consists in the reduction of given forces apparently diverse to a smaller number of forces and powers sufficient for the explication of the former"[33]. This view was important at the beginning of the 19th century, and was related to the German Romantic movement. It had a great influence in biology; in physics, it stimulated the search for relationships between different physical phenomena, including gravitation, electricity, and magnetism.

Before the work of William Gilbert, in the 16th century, the identification of electric and magnetic attractions was common; *De Magnete* clearly separated these effects. However, the fact that amber attracted pieces of paper, in an apparently analogous way to the attraction of iron by a magnet, suggested to many people that there existed something in common, or a link, between the two classes of phenomena.

Some isolated facts documented before the 19th century indeed pointed to this link between electricity and magnetism. One of them was the "strange effect of Thunder upon a Magnetick Sea-card", reported in 1676[34]. A ship hit by lightning at the latitude of the Bermudas had her foremast broken; she then started to sail in the opposite direction! When the captain of an accompanying vessel changed course and reached the damaged ship he verified, in the words of the report, that "the card was turned round, the North and South points having changed positions".

Another report informed that some cutlery became magnetized when its container was struck by lightning in Wakefield, England in the year 1731; the discharge had even melted some knives and forks. When the knives were put on a table near some nails, these were attracted. A certain Dr. Cookson who witnessed this strange phenomenon speculated that it might have resulted from cooling the molten cutlery in the Earth's field [35].

The interest and the continued investigation on the unity of electricity and magnetism ushered in a discovery in 1820 that was to represent a major milestone in science. This was the result of the investigation of a Danish physicist, Hans Christian Oersted (1777–1851), on the connections between these two classes of phenomena.

The Little Hans Christian and the Great Hans Christian

One of the most important sights in the city of Copenhagen is the bronze statue of the Little Mermaid, at the entrance of the port. The Little Mermaid is a creation of the Danish writer Hans Christian Andersen, who lived in the first half of the 19th century and wrote the famous Tales of Andersen. He is regarded as the founder of modern children's literature, creating an innovative language and producing masterpieces that appealed both to children and adults. Andersen is the author of *The Ugly Duckling, The Emperor's New Clothes, The Red Shoes,* among other books.

This genre had been pioneered in the 17th century with the collection known as the *Tales of Mother Goose*, by the French Charles Perrault (1628–1703); this included classics such as *Cinderella* and *The Sleeping Beauty*. In the 19th century, Andersen had been preceded by the German brothers Jacob and Wilhelm Grimm (1785–1863 and 1786–1859), authors of *Hansel and Gretel*, and *Snow White and the Seven Dwarfs*.

Figure 3.4 The Danish scientist Hans Christian Oersted (1777–1851), discoverer of the effect of an electric current on a compass needle.

A contemporary, compatriot and namesake of Andersen, Hans Christian Oersted (1777–1851) (Figure 3.4), left us no fairy tales, but instead demonstrated beyond doubt the intimate connection between electricity and magnetism. When the future writer Hans Christian Andersen arrived in Copenhagen, he was a poor boy of 14; he met Oersted and was helped by him, already an adult. They developed a close relationship, and in his recollections Andersen wrote: "... his home became very early a home for me; his children, when they were small, I have played with, seen them grow up and keep their love for me. In his home I have found my eldest and unchanged friends."[36] Once a week Andersen dined at the Oersteds, and continued this routine for 24 years after Hans Christian Oersted's death in 1851. Hans Christian Andersen referred to himself as 'Little Hans Christian', and to Oersted as 'Great Hans Christian'. And great he was, being nowadays remembered as the author of one of the most important experiments in the history of science.

Oersted was born in Rudjoeping, Denmark, in 1777, a son of the owner of an apothecary shop. At the age of eleven he began to work as his father's assistant in the pharmacy; from this experience, he derived a practical knowledge in chemistry and an interest in the sciences in general. In 1794, the family moved to Copenhagen, and after three years Oersted obtained a degree in pharmacy at the university.

Hans Christian Oersted had a strong interest in philosophy and became a follower of Kantian ideas. In 1799, he obtained the degree of doctor of philosophy, with a thesis on Kant, *Dissertatio de forma Methaphysices elementaris naturae externae* ('Dissertation on the Elementary Metaphysical Forms of External Natures'). This reflected the influence of Kant's ideas expressed in the *Critique of Pure Reason* and in the *Metaphysical Foundations of Natural Science*[37]. Oersted was a member of the editorial staff of the journal Philosophisk Repertorium for Faedrelandets Nyeste Litteratur, devoted to Kantian philosophy.

In the summer of 1801, Oersted started a tour through Europe, visiting several academic and scientific institutions. He visited laboratories in Berlin, Göttingen and Weimar. He attended lectures on *Naturphilosophie* in Berlin, and met some philosophers of that movement, including Johann Fichte (1762–1814) and August Schlegel (1767–1845). Although influenced by these ideas, Oersted remained closer to Kant's system, and was critical in relation to Natural Philosophy for its speculative attitude towards research; he valued the writings of its followers for their 'great beauty', but criticized them for underestimating the importance of observation and experimentation[38]. However, the relevance of Naturphilosophie as an inspiration for Oersted should not be underestimated[39].

Like many other scientists, Oersted interpreted Kant's ideas as implying that the patterns or laws found in natural phenomena were in some sense imposed by the human mind; since God had made man to His image, human reason somehow corresponded to the intentions of the Creator. This idea is vividly expressed by the 20th century physicist Arthur Eddington (1882–1944), in the words he used to describe man's quest for understanding nature[40]: "We have found a strange footprint on the shores of the unknown. We have devised profound theories, one after another, to account for its origin. At last, we have succeeded in reconstructing the creature that made the footprint. And Lo! It is our own."

Finally, Oersted returned to Denmark in 1804 and was appointed professor of physics and chemistry at the University of Copenhagen in 1806, and full professor (*professor ordinarius*) in 1817. He started his investigations on the interrelation of electricity and magnetism in 1813. At that time he thought that by confining an electric current to a thin wire, he could produce heat, and narrowing the wire still further, he would obtain magnetism[41].

While in Weimar, Oersted met the German chemist Johann Wilhelm Ritter (1776–1810), the discoverer of ultraviolet rays and of many important physical phenomena. Side by side with his scientific interest, Ritter cultivated astrology, and tried to correlate the maximum inclination of the ecliptic with the occurrence of discoveries in electricity. In a letter to Oersted, he predicted another outstanding discovery in the last third of the year 1819 or in 1820[42]. Indeed, in 1820 Oersted discovered that when a compass was approached to a wire carrying an electric current, the compass needle moved. The motion of the needle indicated that a magnetic field had appeared perpendicular to the direction of the wire, aligned tangentially to a circle around it. This observation was a turning point in the history of electricity and magnetism: these two distinct and mysterious areas of inquiry were intimately connected!

According to a letter written by one of his associates, Christopher Hansteen, Oersted had observed this effect in a classroom demonstration, by accident. However, this version of the discovery is disputed nowadays, based on the fact that Oersted had been studying the connection between electricity and magnetism for several years[43]. It is interesting to note that, guided by *a priori* expectations on the symmetry of cause and effect, Oersted had apparently tried for several years to observe the effect of electric currents by arranging the compass needle perpendicular to the conductor. However, only when he set the needle parallel to it he did succeed in producing a noticeable effect*.

The observed effect was described by Oersted in the following words[44]: "The magnetical needle, although included in a box, was disturbed; but as the effect was very feeble, and must, before its law was discovered, seem very irregular, the experiment made no strong impression on the audience." Oersted communicated his discovery in the

* This account of how the discovery happened is reinforced by corrections pasted by Oersted onto the manuscript of the text sent for publication in "The Edinburgh Encyclopaedia" (1830). Within a text containing very few corrections, Oersted had written three versions of his account of the discovery, before settling to a final and stronger fourth version, where he stated that the observed results were "strictly connected with his [Oersted's] other ideas". S. L. Altmann, *Icons and Symmetries*, Oxford University Press, Oxford, 1992, p. 40.

same year, in the Annals of Philosophy, in the paper *Experimenta circa effectum conflictus electrici in acum magneticam* ('Experiments on the effects of electric conflict on a magnetic needle'). Describing the interaction between the wire and the compass, he called the "effect which takes place in this conductor and in the surrounding space ... *conflict of electricity*"[45], a denomination that was not adopted by other authors. Oersted visualized the magnetic interaction occupying the space around the wire[46]:

> "It is sufficiently evident from the preceding facts that the electric conflict is not confined to the conductor, but dispersed pretty widely in the circumjacent space."

This work had immediate repercussions all over the world, and Oersted became a well-known figure in scientific circles. In 1824, he founded the Danish Society for the Promotion of Natural Science, and in 1829 became director of the Polytechnic Institute in Copenhagen.

What Hans Christian Oersted had observed was the fact that an electric current creates (or has associated with it) a magnetic field in its vicinity. This field has an intensity that is directly proportional to the intensity of the current, and points along a direction perpendicular to the current flow. The field is everywhere tangential to circles in the planes perpendicular to the conductor of electricity.

Oersted's experiment proved for the first time beyond any doubt the connection of electricity and magnetism. This was also the first experiment of conversion of electric energy (stored in the pile) into mechanical energy – the motion of the compass needle.

In his last work, *The Soul in Nature*, which Hans Christian Oersted left unfinished when he died in 1851, he proclaimed his faith in the unity of the universe: "Spirit and nature are one, viewed under two different aspects. Thus we cease to wonder at their harmony".[47]

Magnetism and Electric Currents

Hans Christian Oersted's discovery produced a wave of excitement among researchers all over Europe. On 4 September 1820, the French physicist Dominique Arago (1786–1853) announced it at a meeting of the Académie des Sciences in Paris. Arago initially did not believe the result, but later, on the 11 September, was able to repeat Oersted's experiment. Stimulated by the announcement of Oersted's discovery,

Figure 3.5 André Marie Ampère (1775–1836), French physicist who made important contributions to the study of electricity and magnetism.

another French physicist, André Marie Ampère (1775–1836) (Figure 3.5), started a series of experiments on the relation of electric currents and magnetic fields, reporting his first observations within a few weeks of Arago's announcement. Ampère's findings would put his name in the pantheon of the founders of the science of magnetism.

Ampère was born in 1775, in Poleymieux, near Lyon, the son of a wealthy merchant, and his life was marked by the social upheaval France was going through; in 1793, with the capture of Lyon by the Republican army, his father was guillotined. Ampère was a man of wide interests: in 1819, he taught philosophy, and in the next year became Assistant Professor of Astronomy at the University of Paris. In 1824, he was appointed Professor of Experimental Physics at the Collège de France. His scientific work was also wide ranging, and included a classification of the chemical elements, published in 1816, that partly anticipated the work of the Russian Dmitri Mendeleyev (or Mendeleev) (1834–1907) on the periodic table of elements. His most important contribution, however, had to do with the relations between electricity and magnetism. Based on his experiments on the motion of the compass needle under the action of electric currents, he invented an instrument to measure the intensity of these currents, the galvanometer, ancestor of the modern standard bench instruments, the voltmeters and ammeters.

Between 5 and 17 September 1820, Ampère repeated Oersted's experiment and found that the compass needle points precisely in a direction perpendicular to the current-carrying wire, when he compensated for the effect of the Earth's magnetic field. He also discovered that the current flows through every component of the electric circuit, including the voltaic pile[48]. Ampère presented a series of papers in 1820 at the Académie des Sciences on the subject of the interplay of electricity and magnetism. On 18 and 25 September 1820, he reported his observations, and put forward his hypothesis on the relationship between the two classes of phenomena. He observed that parallel wires through which electricity flowed interacted with one another: if the current flowed in the same direction, the wires were attracted, if the directions were opposite, they were repelled. He demonstrated this finding on 9 October at the Académie.

There is some difficulty in establishing the correct chronology of Ampère's discoveries, for two reasons: in the first place, because the text of the Memoirs of the Académie does not coincide with the content of the oral presentation, and second, due to the somewhat confused form of his communications[49]. This lack of clarity was recognized by Ampère himself, and was also witnessed by Hans Christian Oersted when the two scientists met in Paris, in 1823. Oersted reported in harsh words his impression of Ampère: "He is dreadfully confused and is equally unskilled as an experimenter and as a debater"[50].

Throughout his researches, Ampère always aimed at the mathematical formulation that described the physical phenomena connecting electric charges and magnetism. He derived a mathematical expression relating the forces applied between two wires that carried electric currents to the intensity of these currents, and to the distance between the wires. Ampère's most important work, *Mémoire sur la Théorie Mathématique des Phénomènes Electrodynamiques Uniquement Déduites de l'Expérience* ('Note on the Mathematical Theory of Electrodynamic Phenomena Deduced Solely from Experiment'), was published in 1827.

Ampère showed that a coil made of copper wire, if free to turn, behaved in the same way as a compass needle. This confirmed the ideas that had occurred to him when he began his investigation[51], and led him to propose the bold hypothesis that the magnetism of the lodestone, or of the needle of the magnetic compass, arose from electric currents flowing inside matter:

> "... from which it follows that a magnet should be considered an assemblage of electric currents, which all flow in planes perpendicular to its axis, directed in such a way that the southern pole of the magnet, which is turned towards the north, is to the right of these currents while always to the left of a current placed outside the magnet, and which faces it in a parallel direction;"[52]

The existence of this relationship of magnetism of matter with electric currents meant that the connection between electricity and magnetism was even more fundamental than the production of magnetic fields by electric currents in the Oersted experiment had demonstrated. The magnetism of the lodestone itself, and of any magnetized body, had its origin in electric currents that circulated within matter.

However, one paradox remained: in a magnet, the electric current required to explain its magnetism would heat it, but magnets were not known to be warmer than other objects! In fact, this objection against Ampère's first results was pointed out very early by another French physicist, Augustin-Jean Fresnel (1788–1827)[53]. The solution to this puzzle was given by Ampère, at the end of 1820 and in 1821 (following a suggestion from the same Fresnel), when he explained that the currents responsible for the magnetism of matter were microscopic ('molecular'), rather than macroscopic, as the usual currents through conductors. He imagined a magnetic material composed of molecules, and in each molecule a coil where an electric current circulated. This idea anticipated the model for the atom proposed by the Danish physicist Niels Bohr (1885–1962) in 1913, which assumed that the positive atomic nucleus is surrounded by the electrons that circulate around it. This atomic or molecular current would not produce the heating effects expected from ordinary, macroscopic currents.

The same explanation was imagined by Ampère to apply to the Earth's magnetic field: this field would arise from electrical currents flowing along the equator; this was the first attempt to discuss Earth's magnetism that had a sound scientific basis.

For his work on electricity and magnetism, Ampère was honored in the naming of the unit of electric current: the ampere, defined as the electric current that produces a given force between two parallel wires. The ampere is one of the seven basic units of the International System of Units (SI), in use today through the Convention of the Meter established in 1960, and signed by most countries. The law that relates the magnetic field to the currents that produce it is also named after Ampère; it is one of the fundamental laws of electromagnetism (see Chapter 4).

The far-reaching consequences of Oersted's discovery of the generation of magnetic fields by currents pushed many investigators to search for the inverse phenomenon: electric effects produced by magnetic fields. Although such an inverse effect was expected, the efforts to observe it were mostly frustrated; in the words of the English physicist Michael Faraday (1791–1867),

> "... still it appeared very extraordinary, that as every electric current was accompanied by a corresponding intensity of magnetic action at right angles to the current, good conductors of electricity, when placed within the sphere of this action, should not have any current induced through them, or some sensible effect produced equivalent in force to such a current."[54]

And he concludes:

> "These considerations, with their consequences, the hope of obtaining electricity from ordinary magnetism, have stimulated me at various times to investigate experimentally the inductive effect of electric currents."[55]

Michael Faraday (Figure 3.6) was born in Newington (now Southwark, London), England, in 1791. His father, a blacksmith, had moved with his family to look for work in London. Faraday had very little formal education, and lived through material hardship; he worked as an errand boy, and at the age of 14 entered a bookbinder's shop as an

Figure 3.6 Michael Faraday (1791–1867), English scientist who made fundamental discoveries on electricity and magnetism; he also invented the electric motor and the dynamo.

apprentice. He then used every chance he had to read about almost every subject; his interest in science was awakened when he read the article 'Electricity' in a volume of the Encyclopaedia Britannica that he was rebinding[56]. In 1810, he started to attend lectures on physics and chemistry at the City Philosophical Society; in one of these lectures, Faraday saw a voltaic pile for the first time.

One of the clients of the bookbinder gave Faraday tickets to the popular science lectures of Humphry Davy (1778–1829), at the Royal Institution. Davy was a Professor of Chemistry and was famous for his excellent public lectures. In October 1812, Davy was temporarily blinded by an explosion that had occurred in the laboratory and therefore needed someone to help him; in the occasion Faraday was accepted by Davy as an amanuensis. Faraday then showed Davy the carefully bound notes of his public lectures that he had attended, and when a position opened the next year, he was hired as an experimental assistant.

In the same year, Davy took Faraday with him as he visited France and Italy, in a tour that lasted over one year, and included many scientific institutions. Faraday then had the opportunity of meeting some of the most important European scientists in different countries. Returning to London, he devoted himself to his experimental studies, mostly in chemistry. In 1821, he started to investigate electromagnetic phenomena, following the stimulus of his friend Richard Phillips (1778–1851). In the same year, he published a paper in the *Quarterly Journal of Science* where he described how a wire turned around a magnet, when electric current flowed through it.

Faraday worked at the Royal Institution for 20 years, and became one of the greatest experimentalists of all times. In his view, the required qualities of the scientist (philosopher) are the following: "The philosopher should be a man [*sic*] willing to listen to every suggestion, but determined to judge for himself. He should not be biased by appearances; have no favorite hypothesis; be of no school; and in doctrine have no master. He should not be a respecter of person, but of things. Truth should be his primary object. If to these qualities he added industry, he may indeed hope to talk within the veil of the temple of nature"[57]. Throughout his long scientific career, most of Faraday's experiments were of exploratory character. In contrast with the ultimate goal of André Marie Ampère, Faraday's aim was not a mathematical description of the electromagnetic phenomena[58]; he rather tried to observe and carefully scrutinize new phenomena, asking in each situation how nature would behave under the given experimental conditions.

Humphry Davy was a close friend of the English poet Samuel Taylor Coleridge (1772–1834), an enthusiastic adept of *Naturphilosophie*; Faraday may have been influenced in this way by these philosophical ideas[59]. Imbued with the belief in the unity of physical phenomena, Faraday made several attempts to observe the inverse of Oersted's discovery, i.e. the generation of electric currents from magnetic fields. He searched for this inductive effect over a period of 11 years. According to his diary, he was finally successful in his trials on August 29, 1831. On that day, he had wound two coils of copper wire around a wooden cylinder, each 100 feet long; the two coils were electrically insulated one from the other, and at the ends of one coil, he connected a battery. To the other coil, he connected an instrument to detect electric currents (a galvanometer). Nothing was detected at the galvanometer while the current flowed through the coil. Nevertheless, a "slight deflection" was noted, but only at the moment the battery was connected, and also when it was disconnected. Faraday also tried the same experiment with two coils wound on an iron ring; the same results were observed, and the intensity of the effect was much enhanced by the presence of the iron core.

In October of the same year he made a very simple experiment that was closer to the realization of the inverse Oersted experiment: moving a magnet in and out through a hollow cylinder on which a coil had been wound, he obtained an electric current flowing through the coil. This current was detected by the deflection of a magnetic needle produced whenever the magnet was moved relative to the coil. This effect was equally observed by moving the coil relative to the magnet.

The phenomena observed in this whole series of experiments can be explained in terms of induced electromotive force (emf) or voltage, in the coil connected to the current detector (either a compass needle or a galvanometer). In the experiment with two coils, the primary coil, connected to the battery, produced a magnetic field at the other (secondary) coil. The time variation of this field induced a voltage in the secondary coil. The same applies to the use of the moving magnet to vary the field, instead of the primary coil. The faster the rate of change of the magnetic field (or magnetic flux) passing through the secondary coil, the larger the inductive effect, i.e. the emf, or voltage. This proved that such time-dependent magnetic fields created an electric field that acted on the electric charges, producing a flow of current. The mathematical expression of this fact is known today as Faraday's Law, and is one of the fundamental equations of Electromagnetism (see Chapter 4).

The experiment with the iron ring is comparable to the experiment with two coils, except that in this case the magnetic field is augmented by the iron magnetization: the secondary coil samples both the magnetic field due to the first coil, and the field due to the magnetization induced in the iron.

Faraday had proved that electric effects could be produced with magnets, but only if the corresponding magnetic fields were varying with time: this is why static experiments had not produced any detectable effect up to that point.

Faraday recognized at once the importance of these findings. He reported them at the Royal Society one month after his observation, and got all the credit for the discovery of this new phenomenon, which became known as electromagnetic induction. In a well-known episode, Queen Victoria (1837–1901) asked Faraday about the utility of induction; "What is the use of a new-born child?" he retorted[60]. The 20th century Czech poet and scientist Miroslav Holub (1923–1998) described the scene in these words[61]:

> When the Queen, over the
> magnetic lines of force
> on Faraday's rough table, asked
> And what use is it?
> Faraday replied,
> gazing lower than her
> lace collar:
> And what use, Ma'am, is a child?

Other researchers had stumbled upon the phenomenon of induction; the American physicist Joseph Henry (1797–1878) had noticed in 1829 a spark as he switched off power feeding a coil wound on another coil, around an iron core. Henry did not report his findings until 1832 in the *American Journal of Science*, long after the publication of Faraday's paper; he would regret his delay in the submission of his results for the whole life[62]. Ampère too had seen the same effect even earlier, in 1822, but did not investigate it in detail[63]. A curious story is that of the Swiss physicist Jean-Daniel Colladon (1802–1893), who, in 1825, took the extra precaution of mounting his measuring instrument, a galvanometer, in another room, to avoid the effect of electrical disturbances, as he switched the current on and off. Each time he reached the other room to check the galvanometer, however, the effect of the induction had already decayed, and he missed it altogether[64]!

The assembly of two coils wound on an iron ring is the first electrical transformer; feeding the primary coil with an oscillating voltage (usually referred to as AC, or alternating current) produces another AC voltage on the second coil, and the voltage in the latter coil is proportional to the ratio of the number of turns in the two coils. Electromotive forces (emf) or voltages can then be stepped up or down, depending on this ratio.

Oersted had found the principle that would lead to the electric motor, and Faraday's discovery of the phenomenon of electromagnetic induction revealed the principle of the dynamo, or electric generator. If one adds the contribution of James Clerk Maxwell (Chapter 4), one has the scientific foundations that would spur the great leap in industrialization in the second half of the 19th century. In this period, for the first time in history, the accomplishment of technical advances required at least some familiarity with the developments that had taken place in the field of the pure sciences[65].

Animal Magnetism

The lodestone has had medical uses since antiquity, either prescribed as a medicine to be ingested, or applied externally to parts of the body, in the latter case used as a source of magnetic field, assumed to be endowed with healing powers. The medical encyclopedia of the botanist Pedanius Dioscorides of Anazarbos (c. AD 40–90), from the 1st century AD recommended its use for "drawing out gross humors"[66]. In the first half of the 16th century, the German-Swiss physician and alchemist Paracelsus (Philippus Aureolus Theophrastus Bombastus von Hohenheim) (1493–1541) employed the lodestone as an ingredient for a plaster used for stab wounds[67]. He rejected the traditional medicine of Avicenna (980–1037) and Galen (c. 131-c. 201), and started a lineage of medical doctors that relied more and more on chemical drugs, rather than on herbs, to treat their patients.

William Gilbert mentions in *De Magnete* authors who have prescribed the lodestone, among those Dioscorides, Galen, Garcias ab Horto. Ever since Gilbert wrote this, many physicians have practiced with magnets; one of the best-known cases was that of the Irish doctor, Valentine Greatrakes (c. 1628-c. 1700)[68] in the 17th century, reported to have effected many 'cures', attracting large numbers of patients[69]. Athanasius Kircher (1601–1680), in *Magnes, sives De Arte Magnetica* ('The Magnet, or About the Magnetic Art') (1641) discusses the healing power of the magnet, and the importance of magnetism in many natural phenomena; he concludes that God is all nature's magnet (*totius naturae magnes*)[70].

In the 18th century, Friedrich Anton Mesmer (1734–1815), a German physician in Vienna, published *De Planetarum Influxu in corpus humanum* ('The Influence of the Planets on the Human Body') (1766), where he exposed his ideas on the influence of the planets on human health. He later heard of experiments by English physicians with magnets and started to think in terms of "animal magnetism", a human faculty that could be restored by the therapeutic use of the magnet. Mesmer stroked with magnets the bodies of people and induced a state one would now describe as equivalent to a hypnotic trance. In 1776, he abandoned the use of the magnet, but still referred to the phenomenon as animal magnetism. He believed that a "magnetic fluid" passed from him to his patients, inducing the trance.

Mesmer was a friend of Wolfgang Amadeus Mozart (1756–1791), who included a reference to mesmerism in one scene of the opera *Così fan tutte* (Thus do they all), of 1790. One of the characters pretends to be Mesmer, and applies a large horseshoe magnet to two ladies, who recover immediately from their illnesses. Also, the first production of *Bastien und Bastienne,* a one-act *singspiel* (type of opera), was said to be held in Mesmer's gardens, in 1768.

Mesmer could hypnotize people without the magnet, and he perfected his technique after he moved to Paris in 1778, adopting the *baquet,* a chest containing acid and equipped with iron parts said to be magnetized. The patients sat in a circle around the chest, as Mesmer made passes over them. He claimed many cures, and in fact he and his followers published hundreds of case histories; with time, his reputation soared. In Paris he benefited from a general atmosphere favorable to science and as a consequence, to therapies that appeared to be scientific[71]. It was said that he had been offered by the French government 20 000 francs to disclose his secrets, but he refused it.

The word 'mesmerism' was employed for some time for the induction of this psychological state in the patients, until it was substituted by 'hypnotism', a term first used by James Braid (c. 1795–1860), a physician in Manchester, in 1843.

Although hypnosis has evolved into a therapeutic tool with wide application, Mesmer's activities found growing opposition from French physicians. This fact led King Louis XVI to appoint in 1784 a commission to investigate Mesmer's practice, formed among others by Benjamin Franklin (1706–1790) and the French chemist Antoine-Laurent Lavoisier (1743–1794): Mesmer was finally accused of quackery and had to leave the country. His use of the *baquet* has survived in the circles of spiritualistic sitters, with no less scientific disrepute.

Mesmer's work had a part in the genealogy of modern psychoanalysis; the itinerary followed was mesmerism, hypnotism and psychoanalysis, and the common concept has been the idea of transference, for Mesmer of a cosmic fluid, for Sigmund Freud (1856–1939) of the creative sources of the unconscious[72].

Further Reading

B. Dibner, *Oersted and the Discovery of Electromagnetism*, Blaisdell Publishing Company, New York, 1962.

S. Sambursky, *Physical Thought: from the Presocratics to the Quantum Physicists*, Pica Press, New York, 1974.

George Sarton, *Introduction to the History of Science*, vol. I, Robert E. Krieger Publishing Company, Malabar, 1927, Reprinted 1975.

A. Wolf, *History of Science, Technology and Philosophy in the 16th and 17th Century*, vol. 1, 2nd edition, George Allen & Unwin, London, 1962.

A. Wolf, *History of Science, Technology and Philosophy in the 18th Century*, vol. 1, 2nd edition, George Allen & Unwin, London, 1962.

Chapter 4

'Acting where it is not': Magnetism and Action at a Distance

"I have no new discovery to bring before you this evening. I must ask you to go over very old ground, and to turn your attention to a question which has been raised again and again ever since men began to think. The question is that of the transmission of force".

James Clerk Maxwell, *in a communication at the Royal Society, in the year 1854*

Magnetic Attraction

The most amazing aspect of magnetism has always been the effect of one object on another without any visible material link between them. This may be observed with a magnet and a piece of iron, or with two magnets. The interaction between two magnets can be simply described through the concept of poles (Chapter 2). In each magnet there are two poles, North and South, somewhat point-like, that exert forces on each other. With like poles the force is repulsive and with opposite poles it is attractive. Therefore, if like poles of two magnets are nearer to each other than the opposite poles, the overall effect will be of repulsion (Figure 4.1a). However, if two magnets are placed far from one another, the forces between like poles and the forces between opposite poles will be approximately the same, and the final result will be neither attraction nor repulsion. In this situation the forces applied on the two poles will in general have equal intensities and opposite directions, and their sum is zero; they may, however, produce a torque. This is the case of the force applied by the Earth's magnetic field on the needle of a compass: the total force is zero (i. e. the needle is not attracted to either Earth pole), but there is a tendency to turn that aligns the needle in the North-South direction. We may then say that in a uniform magnetic field, such as that acting on the needle, there is

From Lodestone to Supermagnets. Alberto P. Guimarães

ISBN: 3-527-40557-7

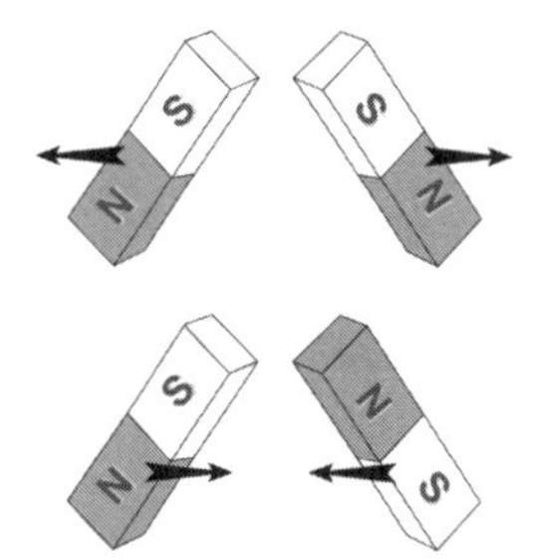

Figure 4.1 Repulsion and attraction between two magnets.

no overall force, and the magnetic attraction that one associates with the action of the magnet is in fact only observed in non-uniform fields, as in the immediate neighborhood of a magnet.

Using two long bar magnets one can measure the forces that one pole of one magnet exerts on another pole of the other magnet. By measuring the force and varying the separation between them, it can be determined how the force varies with distance. This was done by the French physicist Charles-Augustin de Coulomb (1736–1806), who published his results in 1785 in the Mémoires de l'Académie Royale des Sciences; he found that this force is proportional to the inverse of the square of the distance. When the distance doubled, the magnetic force was reduced to one fourth of its original strength. Coulomb also found the same inverse square dependence for the force between electrically charged objects.

As we bring a piece of iron close to the magnet it behaves as if it acquired magnetism. The iron is 'magnetized', that is, it temporarily becomes another magnet, and the force of mutual attraction between the iron and the magnet is of the same type as that between two magnets, as described above (Figure 4.1b).

The phenomena of magnetic attraction and repulsion were explained by most Greek philosophers in antiquity in terms of contiguous action, i. e. action through an intervening medium: the lodestone acted on this medium, which in turn carried the influence of the stone to the piece of iron. The idea of direct action at a distance was in general rejected.

However, with the beginnings of modern science, in the 16th and 17th centuries, the concept of action at a distance gained increasing acceptance, and magnetic attraction became the paradigm of such action. Finally, the idea of action at a distance reached its culmination with Isaac Newton's theory of gravitation.

It was thought, in antiquity in the Western world, that two bodies could only affect one another if they were in direct physical contact, and if this was not the case, something material between them transmitted the influence of one to the other. For instance, for Lucretius (Titus Lucretius Carus), a Latin philosopher and poet who lived in the 1st century BC, the presence of air was instrumental in bringing about the attraction between the magnet and iron: "a stream flows from this stone that pushes the air, creating a vacuum; this attracts the iron"[1]. The great philosopher Plato, who lived in Greece from c. 428 to c. 348 BC, had expressed a similar view in the Timaeus[*2].

Another great Greek philosopher, Aristotle (384–322 BC), thought that objects could only be moved if there was contiguity of source of motion and moved object. He wrote: "There are four ways of being moved by an external agent, namely by pulling or pushing, by carrying or spinning"[3]. Aristotle's conception went one step further in the same direction of attributing importance to the medium, requiring its presence also to maintain the object in motion. Speaking of projectiles, he wrote, "For in the case of these the movement continues even when that which set up the movement is no longer in contact [with the things that are moved]. For that which set them in motion moves a certain portion of air, and this, in turn, being moved, excites motion in another portion; and so, accordingly, it is in this way that [the bodies], whether in air or in liquids, continue moving, until they come to a standstill."[4]

In antiquity the notion of an intervening medium was reinforced by the Stoics, who invoked a *pneuma*, from the Greek for 'air' or 'breath'. An all-pervading mixture of air and fire that had elastic properties, the *pneuma* was able to transmit forces at a distance. An example of an effect at a distance through an intervening medium is given in the writing of Theon of Smirna (c. AD 100), describing the phenomenon of resonating cords: "Strings are in resonance with each other – if on a string instrument one of them is struck, then the other, by some kinship and sympathy, sounds in accord."[5]

* Timaeus speaks: 'the marvels that are observed about the attraction of amber and the Heraclean stones [lodestones], – in none of these cases is there any attraction; but he who investigates rightly, will find that such wonderful phenomena are attributable to the combination of certain conditions – the non-existence of a vacuum, the fact that objects push one another round, and that they change places, passing severally into their proper positions as they are divided or combined.'

Other authors often saw magnetic attraction as arising from a flow of vapors, or 'effluvia', or a stream of corpuscles. This is found in the view of Lucretius, quoted above, and in that of the Neoplatonists. In late antiquity the Neoplatonists attributed effects at a distance to the propagation of corpuscles that could push, or impart a pulling force, the latter in the case of magnetic attraction[6]. Along the same lines, and much earlier, the Greek philosopher Empedocles of Acragas, who lived in the 5th century BC, regarded the action of the magnet on iron (as reported by his contemporary Alexander) as arising from effluences, or vapors, emitted through the pores of both magnet and iron.

The same idea of effluvia survives to the modern age, as shown, for example, in the seventeenth-century writings of the great French philosopher and mathematician René Descartes (1596–1650). Descartes, in his Principles of Philosophy (*Principiorum Philosophiae*), of 1644, attributes the magnetic attraction of iron to the magnet, to 'threaded particles' (*particulae striatae*, in the Latin original) that "expel air between them, making them approach each other"[7]. Two currents of particles flowed through the magnet, one in each direction.

Isaac Newton himself, praised above all as the creator of the theory of Universal Gravitation, hesitated in accepting the possibility of action at a distance. To suppose "that one body may act upon another, at a distance through a vacuum, without the mediation of anything else, (...) is to me so great an absurdity, that I believe no man who has in philosophical matters a competent faculty for thinking, can ever fall into it", he wrote in the year 1692, in a letter to the scholar and priest Richard Bentley (1662–1742)[8].*

In the East, the ideas of action at a distance were more widely accepted, since they fitted better into the framework of Chinese philosophy. This philosophy generally favored reciprocal relations, rather than relationships in a single direction. The difference between the Chinese and the Western view in mechanics would mean emphasis on actions at a distance, over actions of mechanical shock[9].

The Arab philosopher Averroës (Ibn Rushd) (1126–1198) explained the action of the magnet through a mechanism he called "multiplication of the species"; this meant that the lodestone modified the air near it, which in turn modified the air further away, until this influ-

* According to Karl Popper, Newton condemned here, by anticipation, all his followers (Karl R. Popper, *Conjectures and Refutations*, Routledge and Kegan Paul, London, 1972, pg. 107.)

ence arrived at the piece of iron, which then was acted upon by this "virtue".

The knowledge of magnetic phenomena resulted in the establishment of magnetism as the paradigm of action between non-contiguous bodies. This tendency was reinforced with the work of William Gilbert, through his celebrated book *De Magnete*, published in 1600 (see Chapter 2); for Gilbert, magnetism was the determining force that held together the solar system. At that time, many authors used the word 'magnetism' to describe all sorts of attractions, including the gravitational pull. This is seen, for example, in the *Novum Organum*, the major work of the English philosopher and man of letters, Francis Bacon (1561–1626), who, referring to the tides, wrote of "magnetic powers" being responsible for the rise of the waters[10]. The paradigm of magnetism is also clear in the text of the *Epitome of Copernican Astronomy* (1618–21), by Johannes Kepler (1571–1630), on the solar system, where he speaks of similarities and differences between the attraction of the planets by the Sun and the action of the lodestone[11]. And he argues: "But isn't it unbelievable that the celestial bodies be certain huge magnets? Then read the philosophy of magnetism of the Englishman William Gilbert; for in that book, although the author did not believe that the Earth moved among the stars, nevertheless he attributes a magnetic nature to it, by very many arguments, and he teaches that its magnetic threads or filaments extend in straight lines from south to north. Therefore it is by no means absurd or incredible that any one of the primary planets should be what one of the primary planets, namely the Earth, is."[12]

Subtle Matter: the Ether

The idea that the presence of a medium was required to carry interactions between far away objects is very old. The Greek Stoic philosopher Chrysippus (c. 280–207 BC), from Soli, described vision as mediated by a tension in the air between the eye and the object. The original name given to the go-between medium was *pneuma*. In the modern age, from the 17th century, many writers invoked an ether (or aether), from the Greek *aither*, which meant originally air or fire[13]. This would be a medium responsible for carrying the attraction between the planets, or in general, responsible for the influence between non-contiguous objects affected by gravitational or magnetic forces. For the Greeks after Aristotle, and probably for many thinkers before him[14], the ether filled the upper heavens, and this was the stuff stars and planets were made of[15]; above earth, water, air, and fire, the ether was the "fifth element" that constituted the universe.

William Gilbert did his experiments on magnetism at the end of the 16th century. Gilbert and other thinkers, like René Descartes (1596–1650) thought that some fluid matter existing around a magnet had vortices that explained the magnetic attraction. Descartes used the same explanation for the gravitational attraction[16]: "... let us assume that the material of the heaven where the planets are circulates ceaselessly, like a whirlpool with the sun as its centre, and that the parts which are near the sun move more quickly than those which are a certain distance from it". Therefore bodies did not exert forces on each other across empty space; instead, bodies were acted upon through whirlpools in the medium that filled space.

Later, in the 18th century, with the development of mechanistic ideas, magnetic attraction was thought of as resulting from the mechanical properties of an invisible medium. In the same way that a string of iron filings near a magnet transmitted a force from one another, forming the familiar patterns, vacuum also had this property.

As late as in the second half of the 19th century physicists were still devising schemes to explain the role of a material medium as a "cause" of gravity: for example, the Norwegian physicist Carl Anton Bjerknes (1825–1903), searching for a role for this medium, demonstrated in 1874 that an inverse square force of attraction appeared between two pulsating spheres immersed in a fluid.[17]

The requirement of existence of this medium filling space was also postulated to explain the propagation of light. The Dutchman Christiaan Huygens (1629–1695) was one of the proponents of this point of view. Within the mechanistic framework, Huygens, knowing that sound waves propagate more rapidly in less compressible fluids, and also that the speed of light was so immense, concluded that "the particles of the ether to be of a substance as nearly approaching to perfect hardness and possessing a springiness as prompt as we chose".[18] Other thinkers held that the propagation of light required something material, like Descartes, who did not believe in the existence of vacuum: all space was filled with particles – space was a *plenum*. And through these particles, "the action of light is communicated".[19]

Figure 4.2 James Clerk Maxwell (1831–1879), Scottish physicist who first formulated the theory of electromagnetism.

The astronomer Johannes Kepler, born in 1571, also believed in an ether, and thought its existence could be reconciled with the motion of the planets through space. He argued that the ether "yields to the movable bodies no less readily than it yields to the lights of the sun and stars"[20], being so thin that it did not represent an obstacle to the motion of the planets.

Many authors shared this notion that light required a medium as a support for its propagation. Newton also expressed the same view in his treatise on Optics (1704)[21].

Towards the end of the 19th century, James Clerk Maxwell (1831–1879) (Figure 4.2) discussed the propagation of light, initially in terms of stresses in a medium filling space. He later moved away from a purely mechanical picture, refining his ideas in such a way that his ether was regarded by the proponents of fluid models as "too ethereal"[22].

The idea of an ether continued until it was shaken by an experiment on the propagation of light along different directions relative to the motion of the Earth, performed by the German-born American physicist Albert Abraham Michelson (1852–1931) in 1881, and later repeated several times with the assistance of the American chemist Edward Williams Morley (1838–1923) up till 1929. The experiment consisted in the comparison of the time taken by two perpendicular light rays, traveling along paths of the same length in the north-south and east-west directions, made[23] with very precise techniques that used the interference[1] of light. This famous experiment demonstrated that light propagated with the same velocity along the direction of motion of the Earth around the Sun, or perpendicular to this direction. This was not the expected result if the Earth moved relative to an ether filling interplanetary space, where light propagated.

This disagreement had an important impact on the ideas of how light could travel in a vacuum. The final blow that made the idea of ether completely superfluous was provided by Albert Einstein (1879–1955) with his special theory of relativity, published in 1905. No support was necessary for the propagation of light; it was accepted that light (and as we shall see below, all the other electromagnetic radiations) could travel through empty space.

The Apple and the Moon

As we have already mentioned, the idea of action at a distance was fully established with the triumph of the Newtonian gravitation theory. The first ideas of this kind were expressed in more or less precise form by many thinkers, e.g. by the astronomer Johannes Kepler, who understood that two stones, depending on their masses "would come together, after the manner of magnetic bodies".[24] Also, in the writings of Francis Bacon: "Many powers act and take effect only by actual touch, as in the percussion of bodies (....) Other powers act at a distance, though it be very small, of which but few have as yet been noted; although may be more than men suspect" [25].

It was with Newton, however, that the action-at-a-distance point of view reached its culmination. Isaac Newton (Figure 4.3) was born on

* Interference is the effect arising from the superposition of two waves. When two identical waves moving in the same direction meet, and the crests and troughs of them coincide, the effect is that the two waves add up, and the total amplitude is doubled. If, on the other hand, the crests of one wave coincide with the troughs of the other, the effect is a cancellation. In the case of light waves incident on a screen, in the first situation a dark band is produced, and in the second, a light band.

Figure 4.3 Isaac Newton (1642–1727), English physicist and mathematician, one of the greatest scientists of all times.

Christmas Day 1642, near the village of Colsterworth, in Lincolnshire, England. He was a premature baby who grew to become a solitary youth, and entered the University of Cambridge at the age of 18.

In 1665 Cambridge students were sent home to avoid the Great Plague, an epidemic that killed, in London alone, some 70 000 people, which represented about one seventh of London's population. Newton, who had graduated in the same year, returned to the family farm in Woolsthorpe and lived there for almost two years, in a remarkably creative period known as his *anni mirabilis,* the wonder years in the history of science. Many of his discoveries, including calculus, the binomial theorem, the decomposition of white light into the colors of the rainbow, and above all, the foundations of the gravitation theory were made in those years. Fifty years later, he described his achievements in that period with the following words:

"In the beginning of the year 1665 I found the Method of approximating series & the Rule for reducing any dignity [degree] of any Binomial into such a series. The same year in May I found the method of Tangents of Gregory & Slusius, & in November had the direct method of fluxions [differential calculus] & the next year in January had the Theory of Colours & in May following I had entrance into the inverse method of fluxions [integral calculus]. And the same year I began to think of gravity extending to the orb of the Moon & (having found out how to estimate the force with which [a] globe revolving within a sphere presses the surface of the sphere) where Kepler's rule of the periodical times of the Planets being in sesquialterate (i. e. as the power 3/2) proportion of their distances from the center of their Orbs, I deduced that the forces which keep the Planets in their Orbs must [be] reciprocally as the squares of the distances from the centers about which they revolve: & thereby compared the force requisite to keep the Moon in her Orb with the force of gravity at the surface of the earth, & found them answer [agree] pretty nearly."[26] This incredible list of discoveries marks a breathtaking period in the intellectual history of humankind.

The name of Isaac Newton is usually associated with the theory of gravitation through the well-known anecdote according to which Newton had a flash of insight as he watched an apple falling from the tree in his garden. His notes, however, reveal that Newton arrived at the concept of gravitation through a careful analysis, not as the result of a sudden revelation. Revelation or not, the authenticity of an episode with Newton that involved the fall of an apple in Lincolnshire, in the year 1666, is confirmed by at least three independent contemporaries[27]. If gravity reached a branch of an apple tree, being responsible for the fall of the fruit, why wouldn't it reach the Moon, in this case affecting its motion around the Earth? If the Earth did not exist, the Moon would move along a straight line. One could think that at each instant that the Moon deviated from this straight line in its almost circular (elliptical) orbit, it was 'falling' to the Earth, in analogy with the apple.

This was a bold step: to relate the matter-of-fact notion of weight, the tendency of bodies to fall down, to phenomena occurring in the sky. The laws of nature that ruled the sublunar world might extend their kingdom to the realm of the heavens, a concept in complete opposition to the classical Aristotelian view of a world separated into supralunar and sublunar spheres.

The path that led Newton to his discovery of universal gravitation starts from the understanding that the Moon had weight, like the ordinary Earth objects; then he assumed that the same force acted between Sun and planets, and planets and satellites; finally, that this type of interaction was a universal principle, applicable to every object in the universe[28].

Although Newton was reluctant to accept the idea of action at a distance (see above quotation from a letter written in 1692 to Richard Bentley), he came to adopt it as the foundation of his theory of universal gravitation. His profound interest in alchemy made it possible for him to accept this type of action, an element common in the occult sciences*[29].

His view on magnetic attraction was more ambiguous. In the thirty-first Query in the Opticks, he puts together gravity, magnetism and electricity to affirm that "the small Particles of Bodies [have] certain Powers, Virtues, or Forces, by which they act at a distance ... upon one another for producing a great Part of the Phaenomena of Nature."[30] However, it appears that for most of his life Newton regarded magnetic interactions in mechanistic terms, acting through effluvia, in a picture very close to that of most of his contemporaries, and to Descartes'. This is present, for example, in another passage of the second English edition of the Opticks (1717): "And how the effluvia of a magnet can be so rare and subtile as to pass through a plate of glass without any resistance or diminution of their force, and yet so potent as to turn a magnetic needle beyond the glass?"[31] In an unpublished manuscript kept in the Cambridge University Library, he refers to two separate and 'unsociable' 'streams' entering the poles of a magnet[32]. The writings of Newton's followers and close collaborators, including Edmund Halley (1656–1742), John Keill (1671–1721) and Samuel Clarke (1675–1729) also exhibit a nearly 'Cartesian' view of magnetic action[33].

Although recognized in his lifetime as a great scientist, Newton displayed a mean side of his character in many controversies and priority disputes in which he was involved, especially with the German philosopher and mathematician Gottfried Wilhelm Leibniz (1646–1716) on the paternity of the calculus, and with the English physicist Robert Hooke (1635–1703) on the discovery of the inverse-square law of attraction.

Newton's most important published work was the book *Philosophiae*

* "We have every reason to suppose that, had Newton *not* been steeped in alchemical and other magic learning, he would never have proposed forces of attraction and repulsion between bodies as the major feature of his physical system." (J. Henry, 'Newton, Matter and Magic', in *Let Newton Be*, Ed. J. Fauvel, R. Flood, M. Shortland, R. Wilson, Oxford Press, Oxford, 1989, pg. 144.)

Naturalis Principia Mathematica ('Mathematical Principles of Natural Philosophy'), known as the *Principia*, which appeared in 1687, and needed the support and dedication of the astronomer Edmund Halley for publication. The second treatise that summarized Newton's researches was the *Opticks*, published in 1704, at a time when he was the president of the Royal Society. The *Opticks* was written in the form of numbered Queries (16 in the first edition) embodying the results of his discoveries since his time as an undergraduate in Cambridge. In a version of the *Opticks* planned in the early 1690s, Newton had included in Book IV a demonstration of the existence of forces that act at a distance; he now left this out, fearing his critics. In the second English edition, he added Queries 17 to 24 that dealt with the existence of the ether.

Leibniz could not understand why Newton had not tried to find the ultimate cause, or underlying mechanism of gravitation, in his opinion vortices that existed in the ether. Newton spoke of gravitation, or attraction, but did not advance an explanation for its "final cause"[34].

The force of gravity was finally described on a scientific basis within Newtonian mechanics. The idea of action at a distance was applicable to gravitational attraction, as well as to forces between electric charges, and to magnetic attraction and repulsion.

At the end of his life, Newton was acclaimed in England as the most important scientific authority, and founder of a new world order. At the same time, Newton's ideas were not accepted by his contemporaries in continental Europe, where the principle of action at a distance was regarded as archaic, representing a relapse into Aristotelian physics[35]. Mechanical models, which attributed attraction between the heavenly bodies to whirlpools in the ether, were still favored on the continent.

Newton finished the last edition of the *Principia* in 1726, and died in Kensington the next year; on his tomb, an inscription honored him in no mean terms: "Let Mortals rejoice That there existed such and so great an Ornament to the Human Race".

After some time, the winds started to change on the continent: with the growth of empiricism and with the success of Newtonian physics in astronomy, the acceptance of Newton's ideas was finally established[36].

Lines of Force Fill Space

Two centuries after the *Principia*, the English physicist Michael Faraday wrote: "How the magnetic force is transferred through bodies or through space we know not"[37]. Following this statement, he speculated if that interaction was of the same type as the forces between electric charges, whether or not it required the existence of an ether. In his study of electricity, Faraday initially thought in terms of action through contiguous atoms: he was inspired by the way one charged body induces charges of opposite sign on the nearest surface of a neighboring body, which may then produce the same effect on a third body. But with time, his thought tended gradually in the direction of abandoning the idea of action at a distance.

Faraday studied the patterns of the iron filings around the poles of a magnet. The filings form chains that draw curved lines converging to the two opposite poles. The points of convergence coincide with regions of more intense magnetic field. This led Faraday to represent the magnetic field with lines of force [or field lines] drawn in space, such that their direction at each point of space was the same as the direction of the magnetic forces on a piece of iron at that point (or more precisely, as the direction of a compass needle at that point). His concept of lines of force evolved from a mere form of describing the fields, to "physical" lines that mediated the interactions and had all the "reality". *This represented a shift in emphasis that would lead to the concept of field.* "Faraday relegated the particle to the background and enthroned in its stead lines of force throughout space", in the words of one of his biographers[38]. And moreover, Faraday, in his *Experimental Researches* regarded the medium or space around it is as essential as the magnet itself[39].

The stage was set for the birth of the modern concept of field, an idea that had as an early ancestor the *pneuma*, and was also somewhat related to the concept of ether.

The Triumph of the Fields

Before Faraday, the idea of action at a distance had also arisen in the context of the understanding of the properties of solid matter, among them the impenetrability of matter. The Jesuit, Roger Boscovich (1711–

1787) produced, in the 18th century, an approach to the problem of the constitution of matter that brought to the fore the interactions between atoms. For him, the atoms themselves were point-like, and their presence was detected only through their power to attract or to repel. This was also the view of his contemporary, the English clergyman and scientist Joseph Priestley (1733–1804), who argued that solid matter appeared so because of the force of repulsion felt at a distance from the objects*.

James Clerk Maxwell (1831–1879), the outstanding Scottish physicist whose words were quoted in the opening of this chapter, was concerned with the problem of action at a distance; in the same work, in continuation, he wrote[40]: "We see that two bodies at a distance from each other exert a mutual influence on each other's motion. Does this mutual action depend on the existence of some third thing, some medium of communication, occupying the space between the bodies, or do the bodies act on each other immediately, without the intervention of anything else?"

James Clerk Maxwell created his theoretical synthesis upon the foundations set by the experiments of Michael Faraday. Maxwell was born in Edinburgh in 1831, and revealed his aptitude for mathematics very early, winning a prize at the age of 14 with an essay on ovals, which was read before the Royal Society of Edinburgh by a professor of the university. He joined the University of Edinburgh when he was 16, and in 1850 enrolled at the University of Cambridge.

After obtaining his degree, Maxwell stayed for two years at Trinity College, Cambridge, where he read the *Experimental Researches* of Faraday. For some years, he studied the kinetic theory of gases, a subject to which he made important contributions, using a simple picture of molecules as hard particles, like billiard balls. He created the well-known image of "Maxwell's demon", an imaginary being that could sort fast molecules from slow molecules, thereby separating a hot portion (fast particles) from a cold portion (slow particles) of a gas, an

* "To conclude that resistance, on which alone our opinion concerning the solidity or impenetrability of matter is founded, is never occasioned by solid matter, but by something of a very different nature, viz. a power of repulsion always acting at a real and in general an assignable distance from what we call the body itself" (J. Priestley, "Disquisitions Relating to Matter and Spirit" (1777), in *The Science of Matter*, ed. M.P. Crosland, Penguin, Harmondsworth, 1971, p. 116).

enterprise forbidden, under normal circumstances, by the Second Law of Thermodynamics.

In 1860, he became professor at King's College, in London, and for five years dedicated himself to studies of electricity and magnetism. He took over the concept of lines of force from Faraday and tried to formulate a theory that would describe them within a sound mathematical framework. He then produced a model based on the dynamics of fluids, making an analogy between pressure in a hypothetical fluid and electric potential. He showed, in the paper 'On Faraday's Lines of Force', that there was a perfect analogy between the shape of the lines of force of the electrostatic field, and the lines of flow of an incompressible fluid.

Some years later, he further elaborated on the model, imagining vortices that turned around the lines of force, in the paper 'On Physical Lines of Force', still obeying a general mechanical approach. The medium (ether) was divided into cells that turned around the direction of the lines of force in the presence of the magnetic field. To allow neighboring cells to turn in the same sense, he postulated the presence of small spheres that acted as 'idle-wheels' and coupled their rotary motion.

His definitive theory was published in the year 1864, in *A Dynamical Theory of the Electromagnetic Field*; the mechanical images were then reduced to a minimum, and the electromagnetic phenomena were described in terms of fields present on a substrate that had the adequate mechanical properties – the ether.

In 1846, in a paper entitled 'Thoughts on Ray-Vibrations', Faraday had speculated on the possibility of electromagnetic oscillations: "The view which I am so bold as to put forth, considers (...) radiation as a high species of vibrations in the lines of force which are known to connect particles and also masses of matter together. It endeavors to dismiss the ether, but not the vibrations"[41].

In 1855, the German physicists Wilhelm Eduard Weber (1804–1891) and Rudolf Kohlrausch (1809–1858) measured the charge of a capacitor with two different experiments, and derived from the result that the ratio of the values obtained in two different systems of units, electromagnetic and electrostatic, was a number very close to the known velocity of light. This suggested that the electrical disturbances traveled at the speed of light. Maxwell made the daring assumption that light and the electrical disturbances were one and the same thing –

electromagnetic waves. "We can scarcely avoid the inference", he wrote[42].

The remarkable advances reached in the studies of Maxwell can be summarized in four equations, known to this day as "Maxwell's equations", which are regarded as the Magna Carta of Electromagnetism. They embody in quantitative form the description of every phenomenon studied by his predecessor Michael Faraday; these equations connect the electric charges and electric currents to the values of the associated electric and magnetic fields. For static fields, two equations deal with the electric part, and the other two with the magnetic part. However, if the fields vary with time, the intensities of the electric and magnetic terms are related.

Maxwell's equations describe every electromagnetic phenomenon, and even predict the possibility of existence of, and the form of propagation of, electromagnetic waves; they constitute the pinnacle of classical physics.

A few years after Maxwell's death the German physicist Heinrich Rudolf Hertz (1857–1894) detected radio waves, and showed in his experiments in Karlsruhe, between 1885 and 1889, that they had many properties of light waves; they were reflected in the same way, propagated in a straight line, exhibited refraction, and so on. This confirmed experimentally Maxwell's identification of light as electromagnetic radiation. In Hertz's own words[43]: "... this is true for light as such or any special sort of light: of the sun, of a candle, of a glowworm."

The difference between light waves and radio waves resides only in their different wavelengths (or frequencies, with which there is a one-to-one correspondence in vacuum). Waves of light have a length of about one micrometer (0.001 mm), whereas radio waves have wavelengths above one millimeter; for example, FM radio and TV transmissions use waves about three meters long, and AM radio hundreds of meters long. There are electromagnetic waves of much shorter wavelengths, such as X-rays (0.00001 micrometers) or gamma rays (one millionth of a micrometer and below) (Figure 4.4).

But what do we mean when we speak of an electromagnetic wave? It is instructive to make a parallel with a water wave. When a stone is thrown into a lake, it produces water waves that propagate from the point of fall; what we see is a disturbance that moves. A given water molecule on the surface of the lake will be set in motion, as the water wave progresses past it. One may describe this wave as the oscillation

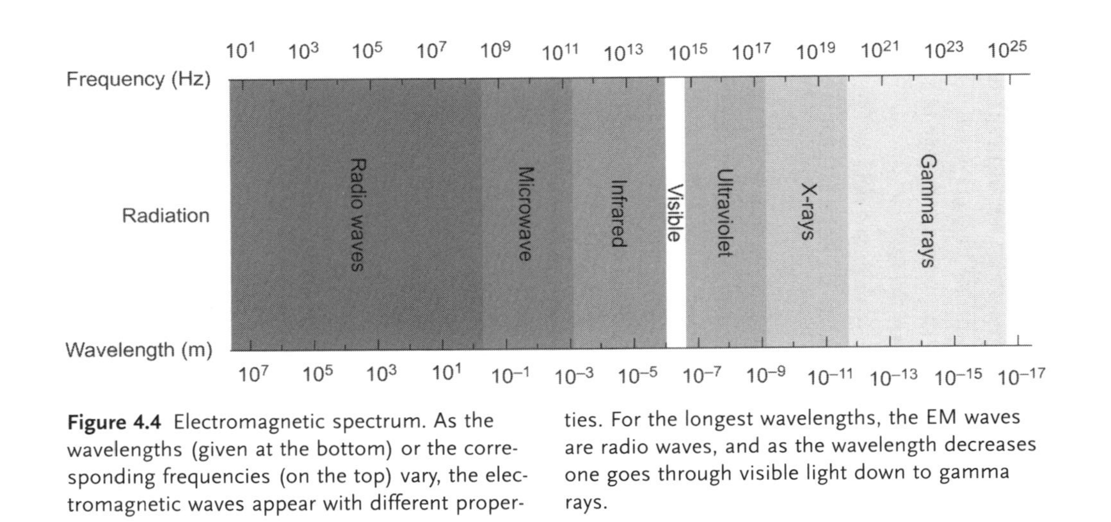

Figure 4.4 Electromagnetic spectrum. As the wavelengths (given at the bottom) or the corresponding frequencies (on the top) vary, the electromagnetic waves appear with different properties. For the longest wavelengths, the EM waves are radio waves, and as the wavelength decreases one goes through visible light down to gamma rays.

in height that affects molecules further and further away from the center of the ripples. An electromagnetic wave also propagates in time, but it needs no material medium. As an electromagnetic wave passes through a given point in the vacuum, what oscillates at that point is the magnitude of the magnetic and electric fields.

Maxwell attributed the inception of the concept of field to Faraday, who had progressed from considering lines of force as geometrical entities to an aspect of physical reality[44].

The concept of field, as well as the idea of its physical reality, i. e. the claim that it was more than a mere mathematical construct, became a cornerstone of Maxwell's theory. This concept made a strong impression on the young Albert Einstein (1879–1955). He wrote in his *Autobiographical Notes*: "The most fascinating subject at the time that I was a student was Maxwell's theory. What made this theory appear revolutionary *was the transition from forces at a distance to fields as fundamental variables* [my emphasis APG]. The incorporation of optics into the theory of electromagnetism (...) – it was like a revelation."[45]

According to the field point of view, one electric charge does not exert a force on another distant charge but, instead, the electric field associated with the first charge acts on the second charge. The field is the agent, not the first charge. We usually say that an electric charge 'creates' an electric field around it, but in fact, the charge and the field are interconnected, and form a whole that cannot be separated.

Physicists give a general definition of field as a continuous distribution in space of values of some physical quantity: one may therefore speak of a field of temperatures in a room, meaning that to each point in space one may attribute a value to the temperature. Another example of field is the ensemble of values of water velocities at different points within a stream of flowing water.

The concept of field is useful when one wants to describe the effects of an electric charge, or of a magnet, in its immediate vicinity. In the neighborhood of an electric charge, for example, another charge will experience a force. The quantity obtained by measuring the force on the second charge and dividing it by the value of this charge is called the magnitude of the electric field. The ensemble of values of this quantity, one for each point in space, is also called the 'electric field' of the first charge.

Albert Einstein and Leopold Infeld (1898–1968), in their *Evolution of Physics*, hailed the concept of field as "the most important invention since Newton's time". And they added, referring specifically to electric and magnetic fields: "It needed great scientific imagination to realize that it is not the charges or particles but the field in the space between the charges and the particles which is essential for the description of physical phenomena"[46]. In Einstein's world view, matter and fields had the same status*.

An interpretation of electromagnetism that opened the way to Einstein's view was that of the Dutch physicist Hendryk Antoon Lorentz (1853–1928). Lorentz tried to follow a middle course between the action-at-a-distance point of view and a 'field' concept. He hypothesized that there was an immobile ether that carried electric and magnetic interactions. To be consistent with this program, he had to move away from some of Newton's postulates; specifically, if charges exerted forces on the ether, it did not apply forces on them, in opposition to Newton's law of action and reaction[47]. He proceeded to derive how the electric and magnetic fields would appear to observers both at rest and in motion relative to an absolute frame of reference (identified with the ether). The results he obtained were essentially the same as those later derived by Einstein in his special relativity theory, except that in Einstein's theory it was shown that the assumption of an ether was not necessary.

Authors tend nowadays to emphasize the conceptual importance of the idea of field, either re-affirming its 'reality', or regarding it as a fundamental mathematical tool for the description of electromagnetic interactions[48]. Some have seen the field, e.g. the electric field, as a property acquired by the space itself in the vicinity of a charge, others as a new entity contained in this space, the space itself remaining simply a geometric container**.

Fields produced by static electric charges and fields produced by magnets are instances of a more general entity – the electromagnetic

* Or in the language of philosophy of science: "Einstein put the field ontologically on a par with matter" (N. J. Nersessian, *Faraday to Einstein: Constructing Meaning in Scientific Theories*, Martinus Nijhoff Publishers, Dordrecht, 1984, p. 35.)

** In the words of the Italian physicist Giuliano Toraldo di Francia (1916–): "... it is also a metaphysical prejudice to allow empty space to have only *geometrical* properties and to discard the possibility that it may have also *physical* properties. It took nearly a century of theoretical and experimental work before physicists began to see this possibility". (G. Toraldo di Francia, *The Investigation of the Physical World*, Cambridge University Press, Cambridge, 1976.)

field. We recall that electric charges in motion generate magnetic fields (Chapter 3). The unity of electric and magnetic fields is more clearly perceived if we compare two situations. Suppose we have an electric charge stationary with respect to an observer; this observer detects the electric field due to the charge. If now the charge begins to move, its presence is gradually perceived by this observer also through its magnetic field. Another observer, on the other hand, travelling with the particle, would always observe only an electric effect. In short, the magnetic and electric parts of the electromagnetic field are manifestations of the same field – electromagnetic – for different observers. This arises from the relativity principle, as formulated by Einstein; although Einstein's relativity theory was created after the findings of Maxwell, Maxwell's equations are consistent with relativity, and in fact they were at the root of Einstein's theorizing.

Another experiment that shows the intimate connection between the electric and magnetic fields is described in Figure 4.5. Let us assume that there are two charged spheres, A and B, with positive and negative charge, respectively, of the same magnitude (Figure 4.5a). Around these spheres, we have an electric field, represented by the field lines, or lines of force. Let us now connect the two spheres with a conducting wire. Since the spheres have either an excess of electrons (sphere B, with negative charge) or a deficit of electrons (sphere A, positive charge), the electrons will start to flow from B to A (Figure 4.5b). Around the wire where the electrons flow, there will appear a magnetic field, and the electric field in the neighborhood of the spheres will gradually disappear. The energy originally stored in the electric field will now begin to appear as magnetic energy. With the continuation of the flow, A and B will reverse the original situation, A will become negative and B will become positive. At a certain point, the flow of charge stops, the magnetic energy vanishes, and the energy appears again in the form of the electric field. These oscillations eventually stop (in a very short time, of the order of a fraction of a second), as the energy is dissipated mostly in the form of heat due to the electrical resistance of the interconnecting wire.

An observer near the two spheres would detect oscillating fields,

Figure 4.5 Two electrically charged spheres connected through a wire, showing electric field lines and the magnetic effect of the electric current (see text).

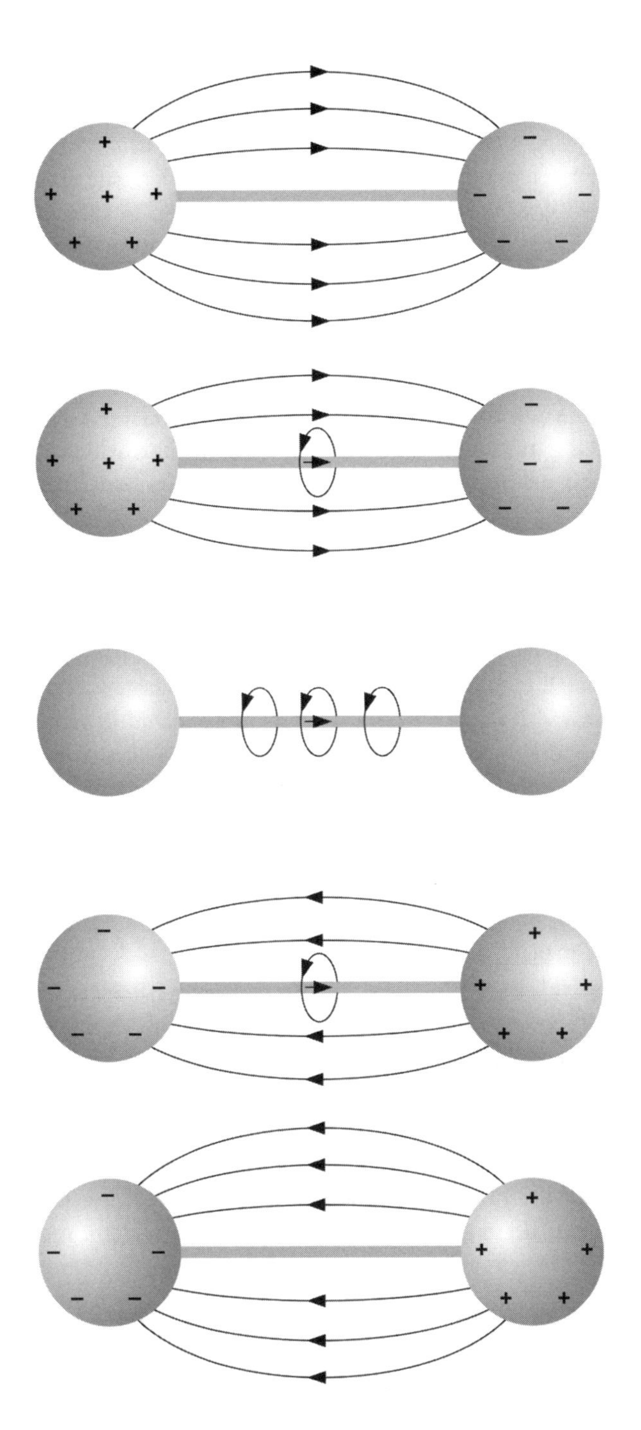

with electric and magnetic components. One way to describe them is to consider that the fields detach themselves from the conductor and charges, and travel in space (Figure 4.5c). As the current in the wire varies, so does the magnetic field near the wire. The time variation of the magnetic field generates (from Maxwell's equations), in turn, a time dependent electric field further away from the wire, and so on. In this way, travelling electric and magnetic fields in space are produced; they constitute the electromagnetic wave. Electromagnetic waves are a general denomination for both radio and light waves; they travel in empty space with a velocity of about 300 000 km/s (186 000 mi/s).

Today physicists know many different fields besides the electromagnetic field and the gravitational field. For example, another field acts inside protons and neutrons, binding together the quarks that form these constituent particles of the atomic nucleus.

We have so far spoken of fields in classical terms: in physics, this means the terms previous to the advent of quantum mechanics, the new physics developed from the 1920s (Chapter 5). In the non-classical descriptions, in the context of quantum electrodynamics (QED), a special theory that incorporated the contributions of quantum mechanics and relativity, developed from the 1950s, the subatomic particles (electrons, muons, quarks, etc) are related to the fields in the sense that they are concentrated bundles, or quanta, of different types of fields. The masses of these particles vary over a very wide range: the heaviest quark – the top quark – is more than 100 000 times heavier than the electron.

The types of interactions between these particles, called 'forces', are also described by fields; for example, the electric and magnetic interactions involve quanta called photons*. These photons participate in the interaction of attraction or repulsion involving charges. Nuclear forces also have their corresponding quanta, in this case called gluons; there are eight types of gluons. All these particles in the interaction fields are virtual particles, which means that they are not directly observable. They are not static; instead, they are continuously being created and annihilated, and the interactions are the result of the exchange of these particles, e.g. between the charges, in the case of the

* A good description of the evolution of this picture of elementary particle physics is given by P.C. Davies and J. Brown in *Superstrings: a Theory of Everything?*, Cambridge University Press, Cambridge, 1992.

electric field. This interaction field interacts with the other type of field, the matter field.

In the QED description of the elementary particle physics there are only fields, and this view embodies, in the words of the American Nobel prize-winning particle physicist Steven Weinberg[49] (1933-), "(...) the central dogma of quantum field theory: *the essential reality is a set of fields* subject to the rules of special relativity and quantum mechanics; (...)"*. The theory that describes these fields is called the Standard Model of particle physics**.

The revolution that started with Michael Faraday one and a half centuries ago led, in the framework of quantum field theory, to a new view of the world, a world of interactions between fields: matter fields and interaction fields. However, most users of magnets do not think in terms of such complex and abstract entities. Their mental picture is different, magnets still interact with iron at a distance, with the interaction somehow represented by the vivid image of the lines of field, imprinted into our minds since Faraday's time.

Further Reading

I. Bernard Cohen, *The Birth of a New Physics*, Penguin, London, 1992.

P.C. Davies and J. Brown, *Superstrings: a Theory of Everything?*, Cambridge University Press, Cambridge, 1992.

Arthur Koestler, *The Sleepwalkers*, Penguin Books, Harmondsworth, 1973.

S. Sambursky, *The Physical World of Late Antiquity*, Princeton University Press, Princeton 1962.

R.S. Westfall, *The Life of Isaac Newton*, Cambridge University Press, Cambridge, 1993.

* "... to field physicists, the fields constitute the fundamental ontology. Spacetime is a structural property of the fields, not the other way around" (S.Y. Auyang, *How is Quantum Field Possible?*, Oxford University Press, New York, 1995, p. 150).

** This theory predicts, to account for the masses of the subatomic particles, the existence of yet another kind of fields, the scalar fields. These fields differ from, e.g. the magnetic field, since the latter has a direction (and is therefore generally described by a vector) while scalar fields do not. The particles of the scalar fields are usually called Higgs particles.

Chapter 5
The Secrets of Matter

"For the first time a plausible story can be told concerning the ultimate magnetic particles, the essential nature of the atom of a ferromagnetic substance, the kind of forces which determine the properties of magnetic crystals (...)"

R. M. Bozorth, in *The Physical Basis of Ferromagnetism* (1940)[1].

Discrete Matter

In what sense does the matter that constitutes the lodestone differ from other substances? What differences in structure account for its weird properties? Although magnetic phenomena have been known for over 3000 years, as we have seen, the understanding of the magnetism of matter, or the comprehension of what makes a magnet different from ordinary matter could only be reached when the atomic constitution of matter was established. Once the existence of the atoms was accepted, one still had to wait for the unveiling of the laws that rule the atomic and subatomic world, discovered in the first decades of the 20th century.

The idea of atoms has a long history, and one of the earliest thinkers to discuss this concept was Democritus (c. 460-c. 370 BC). Democritus, born in Abdera, in Thrace, in what is now the northeastern coast of Greece, was a contemporary of Socrates. He was one of the most influential of the Greek thinkers, author of an extensive work on ethics, natural sciences, literature and mathematics, which unfortunately has not survived to our days. He was a follower of Leucippus (fl. 5th century BC), of whom much less is known. The influence of Democritus is justified since the philosopher Epicurus (341–270 BC) took up many of his ideas and left texts that have been preserved. Although Democritus wrote about so many different subjects, he is mainly remembered in the history of thought as the creator of atomism.

From Lodestone to Supermagnets. Alberto P. Guimarães

ISBN: 3-527-40557-7

The central concept of atomism is that every single object is formed of small immutable particles that cannot be divided (*atomon* – in Greek, that which cannot be cut, indivisible[2]). The Greek philosopher Simplicius of Cilicia (fl. 530 AD), writing in the 6th century AD in *Commentary on the Physics,* stated[3]: "(...) Democritus of Abdera posited the full and the void as first principles, one of which he called being and the other non-being; for he posits the atoms as matter for the things that exist and generates everything else by their differences." To form the material bodies, the atoms stick to one another, since some of them are concave, some convex, some have hooks, and so on[4].

The idea of the discrete nature of matter is found in Lucretius, Roman poet and philosopher of the 1st century BC, a supporter and advocate of atomism. In his *De Rerum Natura* ('On the Nature of Things'), Lucretius explained a series of natural phenomena on an atomistic basis. Thus, he argued that the smallness of the atoms explains why the senses cannot detect the amount of metal that is rubbed each day from a worn ring, or from an iron plowshare[5]. The atoms would confer characteristic properties on different substances in the world around us: for example, things that are pleasant to the senses would be formed of round and smooth atoms[6].

Other thinkers, on the other hand, among them the Stoics (mainly Zeno of Citium (335-c. 263 BC)), believed that matter was infinitely divisible. In the modern era, the seventeenth-century philosopher René Descartes also considered that matter was formed of particles, for example, water was formed of "long, smooth and slippery" particles[7]. However, in his *Principles of Philosophy* (1644) he argues that since extension is an attribute of everything that exists (and therefore occupies space), there can be no indivisible particles, or atoms. Furthermore, matter could in principle be continuously divided, since the opposite would mean that God had deprived Himself of this power[8].

Thus, the question of continuity versus discreteness of matter has been addressed by a variety of thinkers since the time of Democritus, with differing approaches and emphasis. Finally, by the end of the 19th century, the accumulated knowledge on electricity and electromagnetic phenomena, as well as the growing corpus of chemical knowledge, favored a new turn in the discussion of this age-old problem.

'The Real Facts of Nature'

The early pre-Socratic philosophers were the first thinkers to attempt to describe the whole of the universe as formed of a single principle, or element (Chapter 1). As we have seen, this element, for Thales of Miletus, was water. Among other thinkers who suggested primary elements, there was Anaximenes, who suggested air, and Heraclitus, fire. Empedocles of Acraga, or Agrigento, in Sicily (490–430 BC), posited that the multitude of substances in the world around us stem from combinations of the four elements: earth, water, air and fire. This concept was later also adopted by Aristotle, who saw the qualities of the different substances reflecting the proportion of these four elements in their composition. This, in essence, remained the dominant view for 2000 years.

Robert Boyle (1627–1691), English scientist, wrote in *The Sceptical Chymist*, in 1661, against Aristotle's theory of the four elements, arguing that all matter was formed of primary corpuscles, in his words[9], "divided into little particles of several sizes and shapes variously moved." One hundred and twenty years later, in the second half of the 18th century, the Frenchman Antoine-Laurent Lavoisier (1743–1794), one of the founders of modern chemistry, elaborated on the concept of element, giving an operational definition: a substance that could not be analyzed or divided by chemical means. Lavoisier, who also innovated in the use of quantitative techniques in chemistry, established in his *Traité Elémentaire de Chimie* ('Elementary Treatise of Chemistry'), of 1789, that there existed 33 elements.

"An inquiry into the relative weights of the ultimate particles of bodies is a subject, as far as I know, entirely new;" announced the English chemist and physicist John Dalton (1766–1844), in a communication[10] presented at the Literary and Philosophical Society of Manchester, in 1805. And he added: "I have lately been prosecuting this inquiry with remarkable success." The work of John Dalton gave, in the first years of the 19th century, a scientific basis for the belief that atoms were the smallest units of matter. An atom was the smallest portion of a given element that had the same chemical properties as the bulk substance; the smallest portion of a compound was a molecule. Dalton arrived at these conclusions from the study of the proportions in weight of the different elements that entered a chemical reaction; for example, hydrogen and oxygen always participated in the proportion of 2:1 in the chemical reaction that gave rise to water.

In 1869, a Russian professor of chemistry, Dimitri Ivanovich Mendeleyev (or Mendeleev) (1834–1907), presented a chart containing 63 of the then known chemical elements, organized in increasing atomic weights. This table revealed similarities in the chemical properties of the elements and had empty spaces that were associated with elements yet to be discovered. The fact that this chart gave emphasis to several properties of the elements that occurred repeatedly along the series made it known as the 'periodic table'. This table went through many revisions, including a rearrangement according to another characteristic of the elements, their atomic number. This form of presentation represented a major step in the study of the elements, and the scientific importance of the periodic table has been compared to that of Darwin's theory of evolution[11].

Still in the 19th century, studies conducted by Michael Faraday on the chemical decomposition produced by the flow of electricity, and by another English physicist, William Crookes (1832–1919), on electric discharges through a gas at low pressure suggested that there existed negative electric charges.

Joseph John Thomson (1856–1940), a British physicist, wrote: "(...) I can see no escape from the conclusion that they are charges of negative electricity carried by particles of matter. The question next arises, What are these particles?"[12] These particles of negative charge studied by J. J. Thomson were later called 'electrons', and were shown by him to be about 2000 times lighter than the lightest atom.

The electron was the first subatomic particle to be identified, and its discovery represents an enormous advance in the understanding of the constitution of matter. We may also say that this fact inaugurated the discipline of elementary particle physics.

These advances occurred in parallel with progress in the comprehension of the structure of matter, more specifically on the understanding of the fact that matter is made of atoms, and on the nature and constitution of these ultimate building blocks. Perhaps the most important experiment in the quest for understanding the structure of the atom was that performed under the supervision of Ernest Rutherford (1871–1937), a New Zealander physicist who moved to England in 1895. Rutherford had spent some years in Montreal, where he had studied the physical properties of alpha and beta particles, determined the first radioactive half-life, and conceived the radioactive transmutation theory in 1902 (with the English chemist Frederick Soddy (1877–1956)). He returned to England in 1906, this time converting the University of Manchester into a world center for research in physics, and practically creating the discipline that would subsequently be designated 'nuclear physics'. For his remarkable discoveries made in Canada on the radioactive decay of nuclei, Rutherford was awarded the Nobel Prize for chemistry in 1906.

The celebrated experiment was performed in 1909 by Rutherford's assistant Hans Geiger (1882–1945) and a student, Ernest Marsden (1889–1970). It consisted in bombarding a thin foil of gold with alpha particles, which are energetic nuclei of the element helium, emitted in the decay of some radioactive nuclei. It revealed that the majority of alpha particles went right through the atoms, as expected, but a very small proportion of them were reflected, or strongly deviated. "It was quite the most incredible event that has ever happened to me in my life", Rutherford later remarked[13]. "It was almost as incredible as if you fired a 15-inch shell at a piece of tissue paper and it came back and hit you."

Rutherford correctly interpreted this surprising result as meaning that the atom is almost completely void, with a very small and massive nucleus, with a radius 100 000 times smaller than the atomic radius. The electrons, in Rutherford's vivid image[14], whizzed through the empty space around the nucleus, "like a few flies in a cathedral".

Ernest Rutherford was convinced of the primacy of experimental interrogation of the secrets of nature. Speaking[15] of the theoretical physicists, who use as their tools the mathematical description of the physical world, he once said: "They play games with their symbols", and he added, "but we turn out the real facts of Nature". It has been pointed out that Rutherford's was a type of diversifying mind, interested in exploring the multiplicity of natural facts, in opposition to other scientists', like Einstein, who emphasized in their work the search for a unifying picture of the world*.

In 1911, Rutherford met the young Danish physicist Niels Bohr

* The American physicist of British origin John Freeman Dyson (1923-) sees a parallel in these two attitudes, that could be described as 'Baconian' and 'Cartesian', with the cities of (coincidently) Manchester, the first industrial city, and Athens, the first academic city (Freeman Dyson, 'Manchester and Athens', in *Infinite in All Directions*, Penguin Books, London, 1988, p. 35).

(1885–1962) who had graduated at the University of Copenhagen, where he also obtained his PhD, and was then doing postdoctoral work in Cambridge. Bohr was captured by Rutherford's charisma, and decided to come to Manchester to work with him.

Niels Bohr returned to Denmark in 1912, and in the next year published three papers where he proposed the first theory of the atom that involved quantum concepts. He had to postulate that electrons circled the nucleus in stationary orbits, in a picture similar to the solar system. However, in direct opposition to the principles of classical physics, the electronic orbits in his model could only have discrete radii. Classically, any value of orbital radius would be allowed, and the electron would occupy a continuum of energy states, rather than a set of discrete energy values. In addition, according to classical electromagnetism, a charged particle would eventually spiral into the nucleus, and therefore the atom would be unstable.

A significant development that contributed to the knowledge of the structure of matter was the discovery of a mysterious radiation emitted by uranium salts, by the French physicist Antoine Henri Becquerel (1852–1908). The phenomenon of radioactivity, as it became known later, was accidentally stumbled upon as Becquerel noticed in 1896 that a photographic film was exposed in the proximity of the uranium compounds.

The Polish physicist Marie Sklodowska Curie (1867–1934) and her husband Pierre Curie (1859–1906), working at the University of Paris found in 1898 that samples containing the element thorium were also radioactive. After ten years of investigation by the Paris group and by Ernest Rutherford (1871–1937), then at McGill University, in Montreal, it became clear that radioactive atoms emitted three types of rays: alpha-rays, beta-rays and gamma-rays. These would later be identified as helium nuclei, electrons, and energetic photons, respectively.

In 1894, in an often quoted remark, the German-born American physicist Albert Michelson (1852–1931), speaking of the *fin-de-siècle* perspectives of physics concluded: “it seems probable that most of the grand underlying principles have been firmly established and that further advances are to be sought chiefly in the rigorous applications of these principles to all the phenomena which come under our notice”[16]. In a glaring rebuttal of Michelson's appraisal, in an interval of a few years, in the 1890s, some very important discoveries heralded a profound change in physics, with repercussions that would have a very

Figure 5.1 The German physicist Max Planck (1858–1947), founder of the quantum theory.

broad reach, affecting every sphere of human activity. These included the discovery of X-rays (1895), radioactivity (1896), and the electron (1897).

Despite the momentous scientific advances* at the end of the 19th century, the actual constitution of matter was still unknown. The existence of atoms as fundamental building blocks of all matter, and the comprehension of their internal structure, would not be definitely established until the first decades of the 20th century.

The Quantum World

In December 14, 1900, at a meeting of the German (formally the Berlin) Physical Society, Max Planck (1858–1947) (Fig. 1), then pro-

* In some cases, the potentialities of these advances were immediately presumed. For example, a contemporary writing on the observation of X rays described it as "(...) a discovery so strange that its importance cannot yet be measured, its utility be even prophesied, or its ultimate effect upon long-established scientific beliefs be even vaguely foretold." (H.J.W. Dam, McClure's Magazine, April 1896, quoted by Leonard Muldawer in Resource Letter XR-1 on X Rays, American Journal of Physics, February 1969, vol. 37, number 2, p. 124.)

fessor of theoretical physics at the University of Berlin, presented his novel ideas on heat and radiation. He had been studying the frequency spectrum of the radiation emitted by the walls of a furnace, a property that is independent of the material that constitutes these walls. The experimental results that he was analyzing had been obtained at the laboratories of the Physikalisch Technische Reichsanstalt, the imperial bureau of standards in Berlin, some years before. This frequency spectrum, or distribution, is related, for example, to the fact that the color of a heated object changes, as the temperature increases, from a dull red to a bright yellow.

In his work, Planck considered matter composed of atoms, which acted as oscillators (like radio receivers and transmitters) that absorbed and emitted radiation. The novel assumption he made was that the energy of these oscillators assumed discrete values, or energy 'quanta' (plural of the word *quantum*, neuter of the Latin word for how much). This proposal allowed him to explain the frequency distribution of the radiation emitted by heated bodies. It was a revolutionary idea, representing a direct challenge to the prevailing theories, part of the edifice of what came to be known as classical physics, in which the energy of a physical system may assume any value, varying continuously. Although Planck saw this assumption of discreteness, or quantization, as the only way to explain the experimental facts, he was not very comfortable with it, and many years later described[17] "the whole procedure as an act of despair".

Max Karl Ernst Ludwig Planck was born in Kiel, Germany in 1858. He obtained his doctorate in Munich in 1879 on thermodynamics, and became full professor at the University of Berlin in 1892. Besides his qualities as a scientist, he was an excellent musician, and in his youth considered the possibility of following an artistic career; he even composed an operetta, and in later years played in a trio that included Albert Einstein[18].

Planck was for many decades the most important name in the German scientific establishment, "the voice of scientific research"[19], as rector of the Berlin University from 1913, secretary of the Berlin Academy, most influential figure of the Physical Society, and from 1930 president of the Kaiser-Wilhelm-Gesellschaft (Kaiser Wilhelm Society). The latter scientific research organization had been founded in 1911 and in 1948 was renamed the Max Planck Society.

One of the consequences of the original 1900 work of Max Planck

Figure 5.2 Albert Einstein (1879–1955), German-American physicist, celebrated for his great scientific contributions that include the relativity theory and the theory of the photo-electric effect.

was that it gave support to the idea of atomic constitution of matter, at that time still an object of controversy. For the Swedish chemist Svante Arrhenius (1859–1927)[20], this was "the most important offspring" of Planck's proposal.

In the five years following Planck's paper, practically no further progress was made in the direction that he had inaugurated. In 1905 and 1906, however, the then unknown young patent officer in Bern, Albert Einstein (Figure 5.2), applied the same approach to treat two other physical problems. The first was the so-called photoelectric effect, a phenomenon that consists in the appearance of a positive electric charge on some metals, produced by the incidence of light. This originates from the fact that electrons are emitted from the surface of these metals. However, the fact that their energy is directly related to the frequency of the light, but not to its intensity, could not be explained by classical physics. Einstein's proposal took Planck's hypothesis one stage further, assuming that the energy of the light, or electromagnetic radiation in general, was also carried in discrete packets, or quanta. This idea was not at first generally accepted, but would be credited in the end as the correct explanation for the effect.

The other important extension of the new theory, also proposed by Einstein, was the 'quantization' of the thermal vibrations of the atoms in a solid; according to this hypothesis, as a solid was heated, the vibrations of the atoms could have only discrete values of energy. With this suggestion, the relation between the amount of heat absorbed by a body and the increase in its temperature, involving a quantity that physicists call 'specific heat', could be explained for a wide range of temperatures.

Albert Einstein was born in Ulm, Germany, in 1879 to a Jewish middle class family. In his childhood, his family moved to Italy after his father's business failed. Later he settled in Switzerland, where he studied to be a physics teacher at the Swiss Federal Polytechnical School (ETH) in Zurich. After he graduated, he took up the job at the Patent Office in Bern, in 1902. He subsequently occupied different academic positions, remaining from 1914 to 1932 professor of physics at the University of Berlin. Einstein wrote the 1905 papers in Bern practically isolated from other physicists; he once mentioned that the first time he met a professional physicist was when he was 30.

Einstein's contributions to physics include the special relativity theory of 1905, the theory of the Brownian motion, the general relativity theory of 1916, and many important results in the branch of physics called statistical mechanics, which applies statistical methods to different physical phenomena. The general relativity theory dealt with the gravitational field, and predicted that light rays would interact with this field. The observational confirmation of Einstein's prediction was achieved in 1919 as light from the stars was shown to deviate as it passed near the Sun. This success of the theory drew the attention of world public opinion and made Einstein instantly famous.

While in Berlin, Einstein witnessed the growing power of the Nazi party, and felt the asphyxiating climate created by the mounting anti-Semitic campaign, that included a denunciation of 'Jewish science' as unworthy and false. Under such pressure, he left Germany in 1932, finally taking up residence in the USA, and joining the Institute of Advanced Studies in Princeton. There he was destined to play a fundamental role in history, when he signed a letter to President Franklin Delano Roosevelt in 1939 advising that the allies should undertake a project of tapping the energy from nuclei, since there were signals that Germany had already taken this path. This letter helped to set in train a huge scientific, industrial and military machine that would lead to the first nuclear bomb test, in the Alamogordo desert in July 16, 1945*. The ensuing atomic bomb attack on Hiroshima and Nagasaki filled

* Another letter sent much later, this time urging the American government not to use the bomb against Japan, was found unopened on Roosevelt's desk on the day of his death (G. J. Whitrow, Ed., *Einstein, the Man and his Achievement*, Dover, New York, 1967, p. 89).

Einstein with grief, shocked his humanitarian conscience and reinforced his pacifist activism for the rest of his life.

Einstein's direct relation with magnetism, besides the recollection of the impression made by a compass in his childhood (see opening quotation of this book), and a school essay written when he was only 16[21], also includes his only experimental work. This was done in 1915 in the Physikalisch Technische Reichsanstalt, and resulted in the discovery of the so-called Einstein-de Haas Effect. In this work, he tried to find an experimental proof of the electric currents postulated by Ampère, and determined that the currents were real, and that they imparted a torque to a rod of magnetized metal[22].

Albert Einstein became the best-known scientist in the 20th century, and Time magazine chose him in the year 2000 as the most important person of that century[23]. He died in 1955, at the age of 76.

The proposals of Planck, Einstein and others would form the basis of a new physics later to be baptized as quantum mechanics. This name was used for the first time in print[24] in 1914 by Arnold Eucken (1884–1952), editor of the German edition of the annals of the 2nd Solvay Congress, held in Brussels in 1913.

The existence of the light quanta, known as photons, was confirmed in experiments made by the American physicist Arthur Holly Compton (1892–1962) and published in 1922. Compton's experiment showed that photons that had collided with electrons lost some of their energy, and therefore changed their frequency, in agreement with the hypothesis that each quantum of light transported an amount of energy proportional to the frequency of the radiation. This interpretation of the experiment, however, was established only after some years of controversy[25].

In the years 1920–1922 Bohr discussed, based on quantum principles[26], the systematic variation of the properties of the elements of the periodic table. He succeeded in explaining how the chemical and physical properties of the elements varied along the table. The atoms of the elements, starting with hydrogen, with one electron, have increasing number of electrons; he assumed that as each electron was added, occupying a new state, the preceding electrons were not disturbed, remaining in their original states.

The French physicist Louis de Broglie (1892–1987), as part of his doctoral thesis in Paris, published in 1923 and 1924 the hypothesis that ordinary matter exhibited wavelike behavior. Every material body,

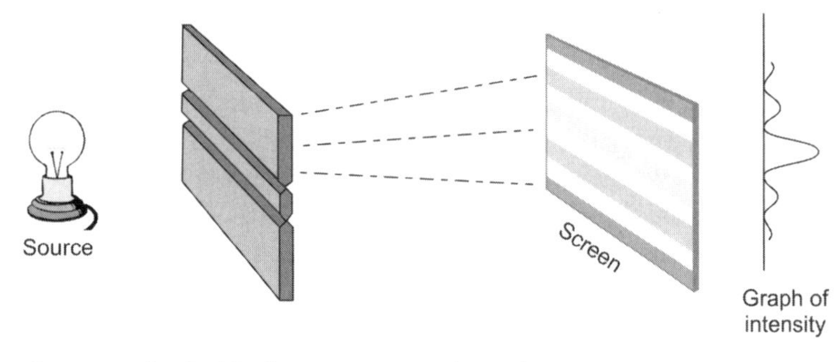

Figure 5.3 The double-slit experiment. Light incident on a screen through two narrow apertures produces a pattern of light and dark lines, due to interference.

of any size, either of atomic or macroscopic scale, behaved as a wave, with a wavelength inversely proportional to its mass. Everyday objects, on the human scale, also have undulatory behavior that we do not detect because their intrinsic wavelength is too small. This daring idea had an experimental confirmation few years later when Clinton Joseph Davisson (1881–1958) and Lester Halbert Germer (1896–1971) in the USA, reported in 1927 that electrons presented wave behavior, and that their wavelength agreed with de Broglie's prediction. These authors had demonstrated that a beam of electrons produced an 'interference' pattern, a succession of light and dark bands on a screen, very similar to that observed with X-rays, known to be electromagnetic waves.

One of the emblematic experiments that revealed dramatically how different the behavior of microscopic or subatomic objects is from that of usual, human-scale objects, is the double slit experiment (Figure 5.3). In this experiment a beam of light is sent through two closely separated openings or slits. Beyond the two slits, a photographic film records the arrival of the beam. On the film, a pattern of dark and light bands appears, a result due to interference, a signature of the wave character of light. This pattern, however, is formed of black dots that arise from the interaction of individual light particles (photons) with the photographic emulsion. Therefore, two physical phenomena, i.e. the interference and the interaction with the film, exhibit, in the first case wave character, and in the second, particle character. Furthermore, the interference pattern is observed even with extremely low intensity of light, that is, even when the photons cross the slits one at a time. Consequently, this pattern is not the result of interference between photons, but one might say instead that it occurs because each photon interferes with itself! The same effect is observed with electrons, and even with atoms.

This puzzling wave-particle duality is a fundamental aspect of nature, which can be detected when one observes photons, electrons, or other particles on the atomic and subatomic scale. Each type of experiment brings to the fore either the particle or the wave character of these entities. According to J. J. Thomson[27], it can be said that in this dual behavior, as in a struggle "between a tiger and a shark, each is supreme in his own element but helpless in that of the other."

The mathematical description of the wave properties of matter was accomplished by another founding father of quantum mechanics, the Austrian physicist Erwin Schrödinger (1887–1961), after he had read de Broglie's thesis in 1925. In a seminar presented the next year in Zurich, where he was a professor of physics, he announced[28]: "My colleague Debye [the Dutch physicist Peter Debye (1884–1966)] suggested that one should have a wave equation; well, I have found one!" Schrödinger's wave equation was applied to a multitude of physical problems, yielding solutions that widened the range of systems whose behavior is determined by the quantum character of matter. This included above all the physics of the atom and the subatomic particles.

The description and understanding of the simplest atom, the atom of hydrogen, was achieved in 1926 by Schrödinger. He was frustrated, however, when the next more complex atom, helium, failed to yield to his equation. Paul Adrian Maurice Dirac (1902–1984), English physicist and another member of the team of founders of the new physics, commented with Schrödinger that this was not important, what really mattered was the aesthetic value of his theory. In a later account of his celebrated remark[29], Dirac maintained that "... it is more important to have beauty in one's equations than to have them fit experiment." Beauty was also more important than the principle of simplicity, or the economy of elements, or parameters, a paramount requirement or criterion in a scientific theory known as Ockham's razor*, a name given in honor of the English Franciscan and logician William of Ockham (c. 1284–1349).

The period from 1925 onwards was characterized by Dirac as the Golden Age of Physics[30]. In the years 1928–1931 he made a major contribution when he generalized Schrödinger's work to make the wave equation compatible with Einstein's theory of relativity. With this modification, another attribute of the electron emerges: the 'spin', analogous, in terms of classical mechanics, to an intrinsic spinning momentum. This had been demonstrated in 1925 by the work of two Dutch physicists, Samuel Goudsmit (1902–1978) and George Uhlenbeck (1900–1988). The spin is an important concept in the understanding of magnetism, as will be discussed in next section.

In the double slit experiment, one cannot predict exactly *where* each individual photon will hit the film, only the probability of its landing on a given region of the film. Another uncertainty shrouds, for example, the decay of a radioactive nucleus. One cannot predict when an individual nucleus will decay; one can only state that after a given time, in an ensemble of nuclei, a certain proportion of them will have decayed. For example, half of them will have disintegrated after a period called the 'half-life'. This impossibility of predicting when the decay of a given nucleus will occur is an essential characteristic of the quantum world, and furthermore, in the words of the German physicist Werner Heisenberg (1901–1976), the "cause for the emission at a given time cannot be found"[31].

Another weird consequence of quantum mechanics is the existence of 'superposition'. This means, for example, that the electron may be in a superposition of states such that its spin points both up and down at the same time! Only when one performs an experiment to determine its state, or when the electron somehow interacts with the medium does it 'choose' which state to show, up or down. This also applies, in principle, to objects of macroscopic scale. Let us take, for example, the experiment of tossing a coin: as the coin falls it is in a superposition of states 'face up' and 'face down'. However, this effect is negligible on this large scale since any interaction, for instance with an air molecule, transforms this superposition practically instantane-

* 'The research worker, in his efforts to express the fundamental laws of Nature in mathematical form, should strive mainly for mathematical beauty. He should still take simplicity into consideration in a subordinate way to beauty ... It often happens that the requirements of simplicity and beauty are the same, but where they clash the latter must take precedence.' P. A. M. Dirac, 'The relation between mathematics and physics', Proc. Roy. Soc. (Edinburgh) 59 (1938/1939) 122–9 (Feb. 25, 1939) (James Scott Prize Lecture), quoted by Helge Kragh, *Dirac – A Scientific Biography*, Cambridge University Press, Cambridge, 1991, p. 277.

ously, defining the outcome of the trial, as in a classical physics experiment[32]. This destruction of this type of so-called coherent superposition of states is known as decoherence.

One of the most remarkable results that came out of the quantum picture is the possibility, in an experiment where two particles are emitted, as the quantum state of one of them is measured, of instantaneously determining the state of the other, irrespective of the distance between them. This is true even when the information from the first particle does not have time to reach the second particle, in an apparent violation of the cherished principle of causality. This experiment was not performed in Einstein's time, but in any case, he did not accept its predicted outcome. He thought[33] that it would mean the result of "spooky actions at a distance". This strange property of nature – often called 'non-locality', nevertheless, has found experimental verification by the French physicist Alain Aspect (1947–) at the University of Paris, in 1982.

This effect arises from a special relationship between states of particles called *entanglement,* which according[34] to Schrödinger is "*the* characteristic trait of quantum mechanics". Due to the property of entanglement, knowing the state of a pair of particles does not mean that the state of each individual particle is known, i.e. in some way the particles are 'inseparable'. It forms the basis of the physical realization of 'teleportation'[35], which can be performed experimentally by passing the state of one particle to another, as demonstrated in the laboratory in 1997.

Quantum mechanics has been extremely successful in the description of, and in making accurate predictions of, phenomena on the subatomic scale. Furthermore, mastering the quantum principles and applying them to the knowledge of the physics of solids has allowed the design and large-scale production of electronic components and devices. It has been estimated[36] that some 30% of the Gross National Product (GNP) of the USA depends on innovations that resulted from the knowledge of quantum mechanics.

Despite these successes, the interpretation of quantum mechanics has been controversial since the very first years. Einstein and Bohr, for example, held a protracted debate during their lifetimes, since they held conflicting points of view on this interpretation. The mathematical equations and the values of physical quantities that are derived in quantum mechanics are not questioned, but how can we understand the counter-intuitive concepts of this remarkable theory? What is the meaning of the fundamental wave function that enters Schrödinger's equation*? One author of several important contributions to quantum physics, the American physicist Richard Feynman (1918–1988) stated bluntly: "[I] think I can safely say that nobody understands quantum mechanics"[37].

Every revolutionary scientific theory stimulates the discussion of the fundamentals of science. Quantum mechanics more than any other theory was conducive to a reassessment of the ideas of causality and determinism, objectivity and subjectivity. The growth of quantum mechanics and the widening acceptance of the validity of its description of atomic and subatomic phenomena stimulated the discussion of the philosophical implications of quantum ideas. It was not only philosophers who participated in this debate, but also many scientists, mostly physicists, took up the challenge of interpreting the consequences of the new physics. "The physicist cannot simply surrender to the philosopher the critical contemplation of the theoretical foundations;" argued Albert Einstein. Moreover, he justified[38]: "for, he himself knows best, and feels more surely where the shoe pinches."

The triumph of quantum mechanical ideas led to a widespread questioning of causality and determinism. Causality is essentially the basic assumption that natural phenomena have causes that precede them. The modern concept of causality had already been expressed[39] by the English philosopher John Stuart Mill (1806–1873) in *A System of Logic*: "The Law of Causation, the recognition of which is the main pillar of inductive science, is but the familiar truth that invariability of succession is found by observation to obtain between every fact in nature and some other fact which has preceded it (...)"

A related idea is that of determinism, that amounts, in simple terms, to the possibility of predicting future events. In its nineteenth-century version, it is embodied in the famous statement by the French savant Pierre-Simon marquis de Laplace (1749–1827), according to whom[40] "An intelligence knowing, at any given instant of time, all forces acting in nature, as well as the momentary positions of all

* Erich Hückel, young physicist at Zurich when Schrödinger presented his wave function expressed his puzzlement in verse:
'Erwin with his psi can do
Calculations quite a few
But one thing has not been seen
Just what does psi really mean?'
(Felix Bloch, Physics Today, vol. 29, December 1976, p. 24).

things of which the universe consists, would be able to comprehend the motions of the largest bodies of the world and those of the smallest atoms in one single formula, provided it were sufficiently powerful to subject all data to analysis; to it, nothing would be uncertain, both future and past would be present before its eyes."

The mechanical determinism exemplified in Laplace's words resulted from the compelling power of Newtonian mechanics in making long-range predictions of many phenomena, particularly in astronomy. The universal scope of these predictions, however, has been denied by the development of deterministic chaos theory in the 20th century, mostly after the 1960s. In this theory, physicists have demonstrated that for a large variety of physical systems, a small uncertainty in the initial conditions leads, after some time, to a broad range of values of the relevant variables (such as position and velocity), making prediction in these cases virtually impossible. Even with phenomena described by simple, well-known physical laws, predictions may be impossible*. Although this is different from the fundamental impossibility derived from quantum mechanics, the practical consequences may be analogous.

One of the pioneers of chaos theory was a meteorologist, Edward Lorenz (1917–), who found that the Earth's atmosphere is an example

* This amplification of effects with time can be illustrated with a passage of a short story by the American writer Ray Bradbury, "*A Sound of Thunder*". Speaking of the consequences of a careless time traveler killing a single mouse in the past, and therefore eliminating the following generations of mice, the guide argues:

"Well, what about the foxes that'll need those mice to survive? For want of ten mice, a fox dies. For want of ten foxes a lion starves. For want of a lion, all manner of insects, vultures, infinite billions of life forms are thrown into chaos and destruction. Eventually it all boils down to this: fifty-nine million years later, a caveman, one of a dozen on the *entire world*, goes hunting wild boar or saber-toothed tiger for food. But you, friend, have *stepped* on all the tigers in that region. By stepping on *one* single mouse. So the caveman starves. And the caveman, please note, is not just *any* expendable man, no! He is an *entire future nation*. From his loins would have sprung ten sons. From *their* loins one hundred sons, and thus onward to a civilization. Destroy this one man, and you destroy a race, a people, an entire history of life. It is comparable to slaying some of Adam's grandchildren. The stomp of your foot, on one mouse, could start an earthquake, the effects of which could shake our earth and destinies down through Time, to their very foundations." (Ray Bradbury, "A Sound of Thunder," in *Classic Stories 1*, Bantam Books, New York, 1990, p. 214).

of a system whose behavior is essentially impossible to forecast. The unpredictable influence of one single factor on the evolution of global weather has been dubbed 'the butterfly effect', after his 1979 paper presented at the American Association for the Advancement of Science, whose provocative title was: *Predictability: Does the Flap of a Butterfly's Wings in Brazil Set Off a Tornado in Texas?*[41] He chose the butterfly inspired by another passage of the same short story by Bradbury quoted in the footnote.

The debate on the philosophical impact of quantum mechanics, in particular on the issues of determinism and causality still goes on. The uncertainty, non-locality and probabilistic aspects of the quantum view seem to be inherent to the physical world, not to limitations in the observer's knowledge. In spite of the great accomplishments of quantum theory, its meaning remains as puzzling as ever. In recent times, some of the counter-intuitive aspects of the quantum world, such as the superposition of states (e.g. electrons with spin both up and down), are regarded, rather than a problem, as promising features that open the way to revolutionary new applications. Among these, one finds the great promises of a hypercomputer using the inherent parallelism of quantum computation, and an unbreakable code using quantum cryptography[42].

Minute Magnets

The advances in atomic theory with the planetary model of the atom proposed by Bohr added a new element to the problem of magnetism. The circulation of electrons around the nucleus represents an internal current on the subatomic scale, that could account for the magnetism of matter. The electric currents that had been proposed by Ampère almost 100 years before were thus available inside the atoms; these contained in fact microscopic current loops, *atoms therefore behaved as tiny magnets*. The understanding that the magnetic properties of matter arose ultimately from the atomic-scale 'magnets' allowed a great advance in the scientific study of magnetism.

In an unpublished work written in 1639–1640, the *Discorso sopra la calamita* ('Discourse on the loadstone') Benedetto Castelli (1577–1643), a disciple of Galileo, argued[43] that every substance had magnetic properties, and could be classified according to whether or not, after being magnetized, it retained its magnetization. He also had the insight that the lodestone was formed of very small magnetic corpuscles[44]: they "are very minute, perhaps to the ultimate degree of smallness, in turning themselves do so in an instant of time, or almost in an instant." This passage is remarkable, since it contains the first suggestion of the existence of magnets of atomic dimensions in the lodestone.

Goudsmit and Uhlenbeck had shown in 1925 that the electrons had, besides their translation motion, an intrinsic angular momentum, analogous to an object that turns around its axis. Because of this intrinsic momentum, known as 'spin', the electron acted as a magnet, that is, it had an associated 'magnetic moment', as this property is called. Since the electron also had a magnetic moment associated with its orbital motion around the nucleus, the electrons had two contributions to their magnetic moments: spin and orbital.

Although atoms have magnetic properties because they contain electrons, inside each orbit in the atom the moments of the different electrons cancel one another when the electron shells or orbits are full, i.e. when they have the maximum number of electrons that they can accommodate. This does not occur in the atoms of the elements with incomplete shells, called transition elements (e.g. iron, cobalt, nickel); consequently, in this case, there is a resulting magnetic moment, and the atom shows paramagnetism (i.e. it will be weakly attracted by a magnet). If the atomic magnetic moments are naturally aligned in parallel, creating a spontaneous magnetization, the material is ferromagnetic (from *ferrum*, iron in Latin).

The advances in the study of magnetism followed two lines: one relied on the moving frontier of the knowledge of the properties of the atom, and another relied on the accumulation of data and on the analysis of overall (or macroscopic) properties of magnetic substances.

In the 19th century, Michael Faraday classified the substances according to their behavior in the presence of an applied magnetic field. He found three main types of materials, according to their magnetic response; these were the diamagnetic, paramagnetic and ferromagnetic materials. The diamagnetic materials or diamagnets exhibited a weak repulsion from a magnet; this property had been discovered by the Dutchman Sebald Justin Brugmans (1763–1819) in 1778, as he noticed that bismuth and antimony were repelled by the poles of a magnetic needle[45]. The paramagnets, e.g. salts of cobalt, showed a weak attraction, and the ferromagnets, like iron metal, a strong attraction. The quantity that measures the magnetic response of a substance is the magnetic susceptibility, given quantitatively by the value of the magnetization in the presence of a magnetic field, divided by the intensity of this field.

The beginning of the modern investigation of magnetism may be identified with a systematic study of the temperature-dependence of the magnetic properties of several substances that was carried out by the French physicist Pierre Curie (1859–1906), who published his results in 1895, in a paper with more than 100 pages[46]. He discovered that many substances presented a susceptibility that decreases as the temperature increases, in an inversely proportional fashion to the temperature in degrees Kelvin, an empirical relationship that was named Curie's law.

For a class of substances that Faraday had called paramagnetic, the susceptibility is positive; for the diamagnetic substances, it is negative, and less dependent on the temperature. Most of the gases were found to be diamagnetic, as well as the liquids and some solids. Most compounds of the transition elements (those with unpaired electrons) are paramagnetic.

Another French investigator, Paul Langevin (1872–1946), proposed, in articles published in 1905 and 1910, a theory that was very successful in explaining the experimental temperature-dependence of the behavior of paramagnets, including the Curie law. Years later, his ideas were confirmed and re-interpreted in the framework of quantum mechanical concepts.

Why are the atomic magnetic moments aligned in parallel in iron metal, and in the other ferromagnets? In 1907, the French physicist Pierre-Ernest Weiss (1865–1940) tried to explain the origin of this ordering of moments: he postulated that to orient a given atom in a piece of iron, the surrounding atoms exert upon it a magnetic field, proportional to the magnetization. No acceptable physical justification was given for the existence of this field, which remained somewhat mysterious until quantum theory was applied to the problem of magnetic ordering. The origin of the effect that aligns the iron atoms was discovered simultaneously by Heisenberg and by Dirac in 1926; it results from the overlap of electrons' states from different atoms. This is an essentially quantum effect, an effective interaction between electrons that mimics a magnetic field. It is related to the fact that in the new physics all the electrons are indistinguishable. This interaction was called 'exchange interaction'. The exchange concept also allowed the understanding of certain chemical forces, for example those that hold together molecules of nitrogen or hydrogen gas.

Weiss was the first investigator to explain why not every piece of iron acts as a magnet, or in other words, why it does not necessarily have a net magnetic moment. Although each small region of a piece of iron has a non-zero magnetic moment resulting from the alignment of the atomic moments mentioned above, normally the total moment of each of these small regions (called 'domains') points along a different direction, adding up to a total moment of approximately zero. In a magnet, on the other hand, although these domains still exist, they point along some preferential directions, and the total sum is not zero. We will discuss in Chapter 7 what materials and what physical or chemical treatments are necessary to achieve this result. In magnetite, of formula Fe_3O_4, the archetypal magnet, although there are moments parallel and anti-parallel within each domain, there is a non-zero total moment for the same reason.

Some samples show magnetic moments per atom that are multiples of an elementary magnetic moment. Weiss made the mistaken hypothesis that this was valid in every case, in an analogous way to the electric charge. For two decades he pursued this idea, searching for it until about 1930[47].

The magnetic properties of a ferromagnet vary with temperature; the spontaneous magnetization decreases with increasing temperature, and finally disappears at a temperature that has been named 'Curie temperature', or 'Curie point'. Iron metal, for example, has a Curie temperature of 770 °C, or 1418 °F. The magnetization of a ferromagnet also changes in a characteristic way when an external magnetic field is applied. The most notable feature of the curve of magnetization versus external field is that the magnetization lags behind the variation of field. It is called the 'hysteresis' curve, from *hysterein*, in Greek 'to be behind' (more in Chapter 7).

The existence of incomplete electronic shells in the atoms provided the correct explanation for the understanding of the magnetism of salts, which are usually electric insulators. The magnetism of metals and alloys, on the other hand, seemed quite different. In particular, the temperature-dependence of the magnetization is different from that observed with insulators. The theoretical explanation for the magnetism of iron, cobalt and nickel and their alloys was developed in the 1930s, with the work of Nevill Mott (1905–1996) and E.C. Stoner (1899–1968), in England, and J.C. Slater (1900–1976), in the USA. These authors used a different approach from that applied to salts, postulating in this case that the electrons occupied bands, i.e. they occupied ranges of energy, as already proposed by the Austrian physicist Wolfgang Pauli (1900–1958) in 1927. In this view, some of the conduction electrons in a magnetic metal have their moments in the 'up' direction (the direction of an applied magnetic field) and some others point in the 'down' direction. The observed magnetic moment, or the magnetization, i.e. the total magnetic moment per volume, is the difference between the total 'up' minus 'down' moment, per volume.

Some new forms of magnetic order, or types of magnetic materials, were added to the original three types (diamagnetic, paramagnetic and ferromagnetic materials). In 1932, the French physicist Louis-Eugène-Félix Néel (1904–2000), and in 1933 the Soviet physicist Lev Davidovich Landau (1908–1968) independently suggested the existence of materials that had atomic magnets of equal magnitude but aligned in anti-parallel fashion – the antiferromagnets. The total magnetic moment of an antiferromagnet is zero. In the same decade, some experimental measurements of magnetization versus applied magnetic field confirmed this prediction, and it turned out that this type of magnetic order is very common. When the atomic magnets are antiparallel, but their moments are not all equal, and therefore there is a non-zero net moment, the material is called 'ferrimagnetic'. Many iron compounds magnetize in this way, among them the iron oxide Fe_3O_4, magnetite, the main constituent of the lodestone.

In 1935 the metal gadolinium, of a family of elements called 'rare earths', was shown to be a ferromagnet like iron. The rare earths include the elements of the lanthanum family, called lanthanides, plus the elements yttrium and zirconium. Two decades later, the very rich magnetic properties of the rare earths were studied in much more detail, with the availability of purer elements.

After World War II, the study of a series of compounds of formula XFe_2O_4 (where X = Ni, Mn, Fe, etc), named ferrites, revealed interesting magnetic properties, which were used in many applications, as permanent magnets and transformer cores, for example.

With the interest in the elements of the uranium family created with the birth of the nuclear age, the physical properties of these elements, called actinides, and their compounds, were investigated. The wider availability of these elements, in larger quantities and with higher purity, has shown that the actinides are also magnetic, and display a strong analogy with the lanthanides.

The study of compounds formed by the combination of metals, mainly those containing the rare earths, showed many interesting magnetic attributes that since the 1970s have been put to the service of the industrial manufacture of permanent magnets.

Most of the studies of magnetism made with metallic samples have used them in the crystalline state, where the atoms are regularly packed side by side. More recently, however, the interest of materials science investigators has turned to the magnetic properties of amorphous alloys, i.e. alloys that are not crystalline. The first amorphous metallic alloys, of gold-silicon, were prepared by the Belgian-born American physicist Pol Duwez (1907–1984) at the California Institute of Technology, in 1959. Amorphous alloys are usually obtained by rapid cooling of molten metallic solutions, so rapid that the metals solidify faster than the crystals can be formed. The amorphous metallic materials, also called metallic glasses, are resistant to corrosion and have found applications that use their high magnetic response. This point will be taken up as we discuss magnetic materials in the 20th century, in Chapter 7.

The study of magnetic metallic films in the 1980s revealed important new results in the physics of low-dimensional solids. Techniques that had been perfected for the manufacture of semiconductor devices allowed the production of magnetic films as thin as one atomic layer. With the techniques of preparation of films, novel materials can be engineered by the superposition of layers of different elements, crystalline or amorphous, each one with the required magnetic characteristics. New structures can be created by arranging layers of atomic thickness, or ultimately, by manipulating individual atoms. This has stimulated the investigation of new magnetic properties and opened the way to many applications, more notably in magnetic sensors and recording heads. Very thin films of magnetic metals exhibit differences in their magnetism compared to bulk samples. Spontaneous magnetic order, for example, usually disappears in films as one prepares thinner and thinner samples, approaching the limit of one atomic layer thickness[48].

Can We Explain Magnetism?

The unveiling of the structure of the atom and the discoveries of quantum mechanics, in particular the idea of the exchange interaction introduced by Werner Heisenberg and Paul Dirac in 1926, explained the mystery of magnetic order in magnetic materials. But what do we really mean when we say that magnetism was 'explained' by these discoveries?

We have seen in earlier chapters how science began and developed from the first efforts of the Greek philosophers, from the pioneering work of the Milesians to the Stoics. This process incorporated different elements of Greek thought, such as the importance attributed to numbers by the Pythagoreans, to the mathematization of the description of nature in Platonism. In addition, it included the development of the principles of logic by the Stoics and Aristotle; and finally, the value of observation by Aristotle and the seeds of experimentation by Archimedes.

The Greek philosophers put forward several ideas to explain magnetism, these "very hard to understand" effects, in the words of the Chinese *Huai Nan Tzu* treatise. These ideas were usually based on mechanistic elements, like those reasonings that involve the flow of air, or the mediation of a vacuum in magnetic attraction. Some descriptions had a vitalistic or animistic character, making the magnet endowed either with life, or with a soul. This could not be otherwise, since the Greeks lacked the knowledge of fundamental physical facts that would only be discovered some 20 centuries later. Essentially, these basic facts concerned on the one hand the structure of matter, and on the other the relationship between magnetic and electric phenomena, in both cases knowledge that would be acquired during the 19th and early 20th centuries. The physics of subatomic phenomena – quantum physics – discovered in the first half of the 20th century, constituted another aspect of the necessary foundation for the explanation of magnetism.

One feels that human beings have an innate urge to 'explain' things; there is a feeling of relief when an explanation is provided for any phenomenon hitherto not understood. A statement that fits an otherwise incomprehensible fact or event into an accepted framework of assumptions, therefore eliminating mystery, is an explanation that produces assurance[49]. "I believe that examination will show that the essence of an explanation consists in reducing a situation to elements with which we are so familiar that we accept them as matter of course so that our curiosity rests", wrote the American physicist and exponent of the operational approach in philosophy of science Percy Williams Bridgman (1882–1961)[50].

What would be the requirements to be fulfilled by a scientific explanation of magnetism? To 'explain' this phenomenon it would be necessary to relate the known empirical facts of magnetism to fundamental physical phenomena, or to general laws of nature. For example, the explanation of gravity was given when Newton established the universal Law of Gravitation, i.e. when he identified as a universal property of matter, the attraction between objects with a force proportional to their masses.

However, philosophers of science disagree on the possibility of science 'explaining' anything at all*. For example, in his celebrated work *The Structure of Science* (1961), the Hungarian-born American philosopher Ernest Nagel (1901–1985) writes:[51] "(...) even if the laws and theories of science are true, they are no more than logically contingent truths about the relations of concomitance or the sequential orders of phenomena. Accordingly, the questions, which the sciences answer, are questions as to how (in what manner or under what circumstances) events happen and things are related. The sciences therefore achieve what are at best only comprehensive and accurate systems of description, not of explanation."

This is not, however, what scientists in general believe and desire to achieve. Scientists are not satisfied until they feel they have discovered what they regard as the 'explanations' for the empirically observed facts. A view with which most scientists would identify themselves is that of W.C. Salmon, who sees scientific explanations as statements that show how the given events "fit into the causal structure of the world"[52]. Or that of the German-born American philosopher Carl Hempel (1905–1997), according to whom an explanation for an event is given both by "(i) particular effects and (ii) uniformities expressed by general laws, from which the event is to be expected"[53]. The latter is very close to the view expressed by the Irish philosopher and bishop

* It seems wiser not to elaborate much on this theme here, since according to R. Torreti, the discussion of the true meaning of "explanation" gave rise to "a vast and most boring literature" (R. Torreti, *The Philosophy of Physics*, Cambridge University Press, Cambridge, 1999, p. 244.)

George Berkeley (1685–1753) some 300 years ago[54]: "(...) explication consists only in showing the conformity any particular phenomenon has to the general laws of nature or, which is the same thing, in discovering the uniformity there is in the production of natural effects". All these three authors, the first two from the 20th century, and the third from three centuries before, agree that science *can* explain the facts of nature. However, they disagree on the importance of causes. The quotations from Berkeley and Hempel are in line with a remark[55] of the French positivist philosopher Auguste Comte (1798–1857), who "proclaimed that explanation by laws, not causes, was the distinctive feature of mature, 'positive' science".

How far can we go when we ask 'why'? For example, we have seen in the dispute between Leibniz and Newton that they disagreed on the need to search for the origin of gravitation (Chapter 4). Can we ask, in the case of gravitation, why matter has this property of attraction?* The answer is that in the modern era it is not regarded as belonging to the province of physics to find answers to these ultimate 'whys'. The Greek thinkers asked many 'whys' that belonged then, and still belong today to the realm of philosophy, or to the branch of philosophy called metaphysics; these were the more general questions like 'What is being?', and so on. Some of these are similar to "questions children ask, but which adult consciousness first dismisses as unanswerable and then forgets"[56]. Many of the questions the early philosophers asked – those more specific – have been answered by the individual sciences; this is the case of many of the questions related to the physical world.

Moving up to another level of still more general 'whys', one can finally ask the question 'Why is there anything, rather than nothing?' This is obviously a very general question; in fact, it is difficult to think of a more all-embracing interrogation than this one, which has been called the super-ultimate question[57]. This issue, maybe the biggest metaphysical problem, has been qualified by many philosophers – perhaps surprisingly, not by all of them – as a meaningless question. Others have described this question as 'extraordinary', or a 'mystery',

* On this question, the English mathematician Karl Pearson (1857–1936) wrote in his classic *The Grammar of Science* that we can "describe how a stone falls to the earth, but not why it does" (*The Grammar of Science*, Everyman, 1937, p. 103, quoted by P. Edwards in "Why", in *The Encyclopedia of Philosophy*, Ed. P. Edwards, Macmillan, New York, 1967, p. 297).

or like the German twentieth-century philosopher Martin Heidegger (1889–1976), a question "incommensurable with any other"[58].

'Extraordinary', or a 'mystery', this puzzling question remains to this day, a very different question from the uncountable 'whys' and 'hows' that scientists ask every day about nature, and from which scientific progress results. And the answers to these queries that relate the problem under scrutiny to fundamental knowledge on the corresponding particular science are for most practical purposes valid explanations.

Further Reading

J. R. Brown, P. C. Davies, J. Broown, Eds., *The Ghost in the Atom: A Discussion of the Mysteries of Quantum Physics*, Cambridge University Press, Cambridge, 1993.

P. Edwards in "Why", in *The Encyclopedia of Philosophy*, Ed. P. Edwards, Macmillan, New York, 1967, p. 297.

Victor Guillemin, *The Story of Quantum Mechanics*, Charles Scribner's Sons, New York, 1968.

J.L. Heilbron, *The Dilemmas of an Upright Man – Max Planck as Spokesman for German Science*, University of California Press, Berkeley, 1986.

Stephen T. Keith and Pierre Quédec, Magnetism and Magnetic Materials, in *Out of the Crystal Maze, Chapters from the History of Solid-State Physics*, Eds. Lillian Hoddeson, Ernest Braun, Jürgen Teichmannn, Spencer Weart, Oxford University Press, New York, 1992.

Helge Kragh, *Quantum Generations, a History of Physics in the Twentieth Century*, Princeton University Press, Princeton, 1999.

Abraham Pais, Introducing Atoms and Their Nuclei, in *Twentieth Century Physics*, vol. I, eds. Laurie M. Brown, Abraham Pais and Brian Pippard, Institute of Physics Publishing, Bristol, 1995.

K.W.H. Stevens, Magnetism, in *Twentieth Century Physics*, vol. II, eds. Laurie M. Brown, Abraham Pais and Brian Pippard, Institute of Physics Publishing, Bristol, 1995.

M. Tegmark and J.A. Wheeler, *100 Years of Quantum Mysteries*, Scientific American February 2001, p. 54.

Chapter 6
Magnets Large and Small

"One influence of the sunspots upon the Earth is perfectly demonstrated. When the spots are numerous, magnetic disturbances ... are most numerous and violent upon the Earth ... The nature and mechanism of the connection is as yet unknown, but of the fact there can be no question."

C. A. Young, *A Textbook of General Astronomy*, Ginn and Co., Boston, 1888[1].

Introduction

From the end of the 16th century when Gilbert attributed the orientation of the compass to the magnetism of the Earth, it became clear that objects of widely different size scales – the magnetic needle and the planet Earth – showed magnetic 'activity'. In modern times it was found out that this scale is much wider, since most subatomic particles, as well as galaxies, are surrounded by magnetic fields.

In the microworld, the magnetism of the particles that are constituents of matter, or of those that are only observed in collisions induced in particle accelerators, is related either to the intrinsic angular momentum, or spin, or to their orbital motion, as is the case, for example, of the electrons moving around the atomic nucleus. The most important particle from the point of view of magnetism is the electron, since, as we have seen, it provides the main contribution to the magnetism of matter. It is also the motion of electrons in a macroscopic scale through wires and coils that creates magnetic fields in electromagnets and solenoids.

Besides the Earth, other astronomical bodies have associated magnetic fields, particularly the Sun and some planets. Our Milky Way and other galaxies also manifest magnetic properties that have been revealed through optical techniques, i. e. techniques that extract the information on magnetic fields through the analysis of the light emitted

From Lodestone to Supermagnets. Alberto P. Guimarães

ISBN: 3-527-40557-7

from these bodies. In particular, spiral galaxies show a pattern of fields that have been described as 'magnetic arms', magnetic field lines that are located between the spiral distribution of stars.

Electrons and galaxies are physical objects that differ in size by a factor of 10^{36} (radius of particle 10^{-15} m, radius of galaxy 10^{21} m), or a trillion trillion trillion times, that is, a number with 36 zeros. This is the fantastic range of scales in which magnetic phenomena are observed.

Magnetic activity is not only experienced in the inanimate world. On our planet, life has evolved immersed in the local magnetic field, and somehow many living creatures sense and make use of this field for orientation.

The Great Magnet

Magnus magnes ipse est globus terrestris – "the Earth globe itself is a great magnet", stated[2] Gilbert in 1600, in the *De Magnete*. This discovery is remarkable in many ways and represents an important step in the knowledge about the Earth; in fact, this magnetic property amounts to the second overall attribute to be associated with our planet, the first one being its roundness[3]. Although it is in fact a great magnet, the Earth is not like a huge piece of magnetite; its magnetism is more like that of an electromagnet, arising from macroscopic electric currents, as will be explained below.

The magnet Earth creates in its neighborhood, and especially on its surface, a magnetic field that has many observable effects; the power to align the magnetic needle was the first of these effects to be discovered (Figure 6.1). The characteristics of this field have been studied by Gilbert and by innumerable researchers in the ensuing 400 years: its intensity, direction relative to the surface of the planet, variation with time, and so on. The magnetic properties of the Earth could only begin to be explained after the relationship of electricity and magnetism had been unveiled, in the first half of the 19th century.

As we have already discussed in Chapter 2, the magnetic field of the Earth in general deviates at each point of the surface from the direction of the meridian – this deviation is called declination. At the end of the 15th century, western European sailors knew that the compass pointed east of true north. In September 1492, the sailors under the

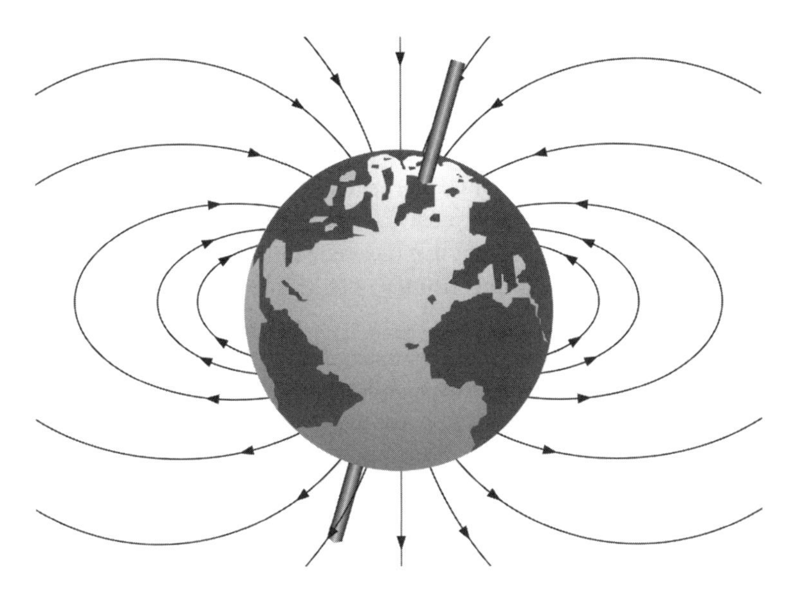

Figure 6.1 Earth magnetic field lines.

command of Christopher Columbus (c.1450–1506) in his first voyage were "afraid" and "dismayed" when, west of the Azores, the magnetic needle shifted to a westward declination. This meant that the declination changed from point to point on the surface of the Earth, and Columbus' entry in his journal on September 13, 1492 is the first report of this westward declination in the western Atlantic[4].

The first worldwide study of declination was performed by the Portuguese naval commander and scholar João de Castro (1500–1548), who sailed to the East Indies and to the Red Sea in 1538, making some 50 observations[5]. He derived the direction of the true north from the position of the Sun, and compared it with the direction of the magnetic compass, thereby deriving the declination at each point. Castro was later appointed viceroy of Portuguese India, where he died, in Goa, in 1548.

In the following century, in 1698 and 1700, the English astronomer and mathematician Edmund Halley (1656–1742), compiled the first chart of values of declination, with data collected in an expedition that constituted the first sea voyages organized for purely scientific purposes. His *Atlantic Chart* was published in 1701, and it represented the magnetic declination by means of contour lines. This is the first known example of representation of a variable through contour lines, a type of graph subsequently called 'Halleyan' curve[6].

The deviation of the magnetic needle from the horizontal plane – the inclination, or dip, was first studied systematically by Gilbert, who found that the compass pointed horizontally at the equator and turned gradually away from the horizontal plane as one moved towards the poles. Gilbert initially thought that the angle between the compass and the plane gave directly the latitude, but this is only a very rough measure.

A magnetized needle that is displaced from the direction of the magnetic field and then released will tend to re-align itself with the field, oscillating until it reaches parallelism with it. The needle will oscillate more rapidly the stronger the intensity of the field. This fact has been used to measure the intensity of the Earth's field, by deriving it from the period of oscillation. The first survey of magnetic field intensity using this technique was made by the French Robert de Paul de Lamanon (1752–1787), in the expedition led by Jean François Galaup count de La Pérouse (1741-c. 1788)[7], French navigator. Unfortunately, the data were lost with the sinking of his ship *La Boussole*. In 1791–1794, Admiral De Rossel determined the field intensity as a function of latitude, verifying the systematic increase from the equator to the poles; his results were published in 1808. This dependence of the magnitude of the field with latitude was also observed by the German naturalist Alexander von Humboldt (1769–1859) in his scientific mission of 1799–1803. He valued these findings so much that he described this accomplishment as "the most important result" of his American voyage[8].

The geomagnetic field tends to be more intense near the poles, with a maximum of about 6/100 000 teslas (or 0.6 gauss) near the poles, and minima near the equator; the lowest value, of about 2.4/100 000 teslas is found on the South American continent, near Rio de Janeiro, Brazil[9].

The magnetic field of the Earth also exhibits a variation with time, both in intensity and in declination. In 1634 the English astronomer Henry Gellibrand (1597–1636), professor at Gresham College, in London, discovered the variation of declination with time. This effect has since been followed for more than three hundred years. The English watchmaker George Graham (1674–1751), pointing a microscope to the tip of a magnetic compass, detected some very rapid oscillations that occur in a matter of seconds, due to very small changes in declination.

Variations of the intensity of the magnetic field as a function of time are observed on different time scales. There are changes that occur on a very short timescale, in the range of thousandths of seconds, and there is a slow variation, on the scale of millions of years. On the long time scale, it is known that the intensity of the field has been decreasing during the last thousands of years at a rate of about 5% per century, and is therefore expected to reach a value of zero around the year 4000, if this tendency persists.

The direction of magnetization of the planet Earth has also changed many times in the past; in the last six million years it has reversed approximately once every 300 000 years.

The knowledge of the long-term variations of the Earth's magnetic field is mainly based on the study of magnetic properties of rocks. The principle is simple: igneous rocks, formed, for example, from lava flow, if solidified under a magnetic field will show some residual magnetization, and this can be related both in intensity and direction to the extant magnetic field. Sedimentary rocks, formed under the same condition of magnetic field, also retain a magnetization. This magnetization is called in general remanent magnetization (thermoremanent, in the first case). Another mechanism of magnetization arises from the enormous electric currents that flow when, during a thunderstorm, lightning reaches the ground.

The Irish physicist Joseph Larmor (1857–1942) proposed, in the 20th century (1919), a theory to explain the origin of the magnetism of the Sun; this is called the dynamo model. The theory was later applied to the Earth, and has been refined by several researchers, remaining the basis of the physical explanation of the Earth's magnetism*. It attributes to the rotating molten iron core of the Earth the role of producing, like a dynamo, an electric current that is responsible for the magnetic field. Long-period changes in the complex motion of the magma account for the variations in the characteristics of the geomagnetic field.

The magnetic field in the neighborhood of the Earth affects the

* The idea is based on the principle that an electrically conducting disc turning in a magnetic field generates a potential difference between its center and its edge, and is therefore a dynamo. If the current produced by this dynamo somehow reinforces this magnetic field, we have a self-exciting dynamo.

motion of charged particles emitted by the Sun (mostly protons and electrons), which constitute the so-called 'solar wind'. The influence of the field effectively traps these particles, creating charged clouds at a distance of thousands of kilometers from the Earth, called Van Allen belts. Explosions in the Sun, called flares, emit large surges of particles that reach the Earth after traveling for about two days. When they reach the upper atmosphere they produce a phenomenon called a magnetic storm, in which the magnetic field measured on the Earth surface is modified and which may lead to the disruption of short-wave communications.

The spectacular display of the *aurora borealis*, or northern lights, was first attributed to the Earth's magnetic field by Edmund Halley in the 17th century[10]; it is known today that the lights arise from the effects of the solar wind particles captured by this field as they hit the outer layers of the atmosphere. The same phenomenon occurs in the southernmost latitudes, and receives the name of *aurora australis*, or southern lights.

The Moon is also a magnet, but much weaker than the Earth; the measure of its strength, the so-called magnetic moment, is one hundred million times smaller than that of our planet. The small resulting magnetic field measured at the Moon's surface is due to remanent rock magnetism[11]. There is no active dynamo mechanism in our satellite. Jupiter and Saturn, on the other hand, have strong magnetic properties, associated with the dynamo effect; Jupiter has a magnetic moment ten thousand times bigger than that of Earth.

Uranus and Neptune have uncommon magnetic properties. Both planets present magnetic moments that are tilted by large angles (of about 50 degrees) in relation to their axes of rotation. The source of magnetism of these planets does not seem to be located near the planet center.

Many other astronomical bodies are magnetically active; the most important example for us is that of the Sun, which produces a field that extends in space over huge distances, reaching the orbit of the Earth, with a value at this distance 10 000 times smaller than the maximum field due to our own planet.

Magnetic fields play a very important part in solar phenomena, the most striking of which is the cycle of sunspots, darker regions that appear this way since they have temperatures more than 1000 degrees lower than the visible surface of the Sun. The invention of the telescope led to the discovery of sunspots, first reported in 1609 by Galileo and other observers. The German pharmacist and amateur astronomer Samuel Heinrich Schwabe (1789–1875) published in 1843 evidence for the existence of a sunspot cycle[12]: every 11 years, the number of sunspots goes through a maximum and decreases again; it was later established that every two 11-year cycles, the north and south magnetic poles change polarity.

The overall magnetism of the Sun arises from essentially the same dynamo mechanism that is active in the Earth, with the extra complexity introduced by the fact that the ionized (electrically charged) gases in the Sun are conductors of electricity. Sunspots exhibit magnetic fields thousands of times stronger than those observed on the Earth surface.

Magnetic fields in stars can be detected through their effects on the emitted light and radio waves; this also allows an estimate of the field intensities. Some stars are known to be relatively small objects, with a radius of about 10 km; these are the neutron stars, stars of ultra-dense matter, formed practically only of neutrons. These strange bodies have very high magnetic fields, of some 10^{10} tesla, or one hundred trillion times (10^{14}) higher than Earth fields. Some neutron stars have an even higher field, the highest field found anywhere in the universe, with a magnitude one thousand trillion times (10^{15}) the Earth field. This class of stars is called 'soft gamma repeaters' (SGR), or *magnetars*. These stars were observed for the first time in 1979, when one SGR produced an ultra-powerful pulse of X-rays, so powerful that in the first two-tenths of a second it emitted more energy than the Sun radiates in 1000 years.

The presence of magnetic fields has also been detected in distant galaxies and even in intergalactic space. The fields in the spiral galaxies tend to be stronger between the spiral arms, and parallel to them, with a low intensity, some 100 000 times smaller than the field at the Earth's surface. Their origin is not completely understood; they are probably due to an amplification of original fields that had arisen in the early moments after the Big Bang. This amplification could result from a dynamo mechanism, as the one that acts in the Earth core; some authors believe that the presence of these original magnetic fields was important for the shaping of galaxies[13].

Electrons and nuclei of high energy – the cosmic rays – impinge regularly on the Earth's high atmosphere. These particles, with velocities approaching that of light, collide with nuclei in the atmosphere and produce showers of secondary particles that can be detected at sea level. The acceleration of ultra-high energy cosmic rays, one of nature's deepest secrets, may be explained by action of magnetic fields outside our own galaxy[14].

From immense galaxies down to a small lodestone, magnetic properties vary enormously in scale; however, these always arise either from the intrinsic moments or from the motion of electrically charged particles.

Living Magnets

The extraordinary ability some birds have of finding their way, either migrating for very long stretches, or returning to their shelter when released a large distance away, has been known for a long time. Carrier pigeons were used by Julius Caesar (c. 100–44 BC) to send to Rome the news of the conquest of Gaul, and by Genghis Khan (c. 1160–1227), in the 13th century, as his army progressed into Eastern Europe. The earliest well-documented case is the use of pigeons for a period of six months during the siege of Paris, during the 1870–1871 Franco-Prussian War[15]. The birds were taken out of Paris by balloons and the messages were photographically reduced; one bird once carried more than 40 000 messages. This use of pigeons continued into the 20th century, during the first and second world wars.

In the 19th century, investigators began the scientific study of migrating and homing birds, trying to determine whether the capacity of these birds of navigating over unknown terrain was related to the use of some internal magnetic compass[16].

Living organisms, since their appearance on our planet some four billion years ago, have existed in environmental conditions that include the presence of the geomagnetic field. Although this field has changed both its intensity and direction on such an extended time scale, it may have played a role in the preservation of life on Earth. This is because the charged particles of the solar wind tend to be trapped in the Earth's field in space, therefore reducing their flow to the surface of the planet. Such reduction probably contributed to the survival of the first living organisms.

One of the earliest studies of the effect of magnetic fields on living creatures was performed[17] in the laboratory of the American inventor Thomas Edison (1847–1931) in 1892: the largest electromagnet existing at that time was used to apply a field of 0.15 T to a boy and a dog, with no apparent effect. (The study concluded that "The ordinary magnets used in medicine have a purely suggestive or psychic effect and would in all probability be quite as useful if made of wood.").

Another line of inquiry has been the attempt to discover the influence that the Earth's magnetic field might have on the behavior of living beings. Is the ability of birds that travel for thousand of miles somehow related to the geomagnetic field? Do they have some form of internal magnetic compass? Through experiments made with applied magnetic fields, data on magnetic orientation, obtained especially since the 1960s, have shown that many living creatures, both invertebrate and vertebrate, have this type of ability; among the latter are fish, amphibians, reptiles, birds and mammals. Investigations with human subjects have found a marginal response to the influence of environmental magnetic fields, and the results are controversial[18].

Many studies have demonstrated that pigeons can find their way using visual cues, using the position of the Sun, as well as information from the geomagnetic field. Under overcast skies, the importance of magnetic data becomes fundamental for their guidance. Experiments have shown that the navigating ability of the birds can be hindered by increases in magnetic activity in the atmosphere, as those occurring during magnetic solar storms. In addition, it has been shown that pigeons with magnets attached to their bodies may be disturbed in their flight back to the coop.

To accomplish their navigation exploits pigeons need to define a direction, i.e. they require what one might call a compass sense, and also determine their position (map sense). The findings point to the use of magnetic compass information by the birds. However, the relevance of magnetic factors for their map building, that makes use of the variation in inclination (or dip) as a function of latitude, and variations in intensity of the magnetic field, is not so well established[19].

In vertebrates two mechanisms of orientation seem to act: in birds and sea turtles, the inclination of the magnetic field is used to define 'poleward' (i.e. the direction towards the pole, indifferently to which pole, north or south) as the direction along which the angle between the vertical and the field direction is smaller. On the other hand, in salmon and mole rats, the internal compass distinguishes north from south, and their navigation uses this knowledge[20].

Honeybees make use of magnetic information in their foraging excursions, and also employ it to align their combs. The behavior of other animals, such as eels and newts, also shows evidence of magnetic orientation.

The simplest living beings that show magnetic orientation are bacteria; this property was discovered in 1975 by accident[21] by R. P. Blakemore, then a graduate student of microbiology at the University of Massachusetts. For these bacteria which were named *Magnetospirillum Magnetotacticum*, the living organisms have a passive role, in the sense that the geomagnetic field orients them so that motion of the bacterial *flagella*, hair-like structure that functions like a propeller, conducts them along the lines of the field. These are called 'magnetotactic' bacteria, from the root *tactic*, that is, directed or oriented by some agent. Since in most regions of the planet the field lines point either up or down, this alignment may help bacteria to avoid the oxygen-rich medium near the surface of the ponds, and search for nutrition near the bottom. In confirmation of this hypothesis, it is observed that magnetic bacteria in the southern and northern hemispheres have opposite polarities.

The discovery of magnetite in these bacteria, where it has the form of chains of grains of 1/10 000 mm (or 1/250 000 in) in diameter, stimulated a search for magnetic substances, or 'compasses' in the tissues of different animals. Many instances of magnetic substances have been found since then in living creatures, such as honeybees, pigeons, turtles and salmon. In practically every case, the magnetic substance is formed of very small grains of magnetite synthesized in the organisms, through a process known as biomineralization. The mechanism through which the interaction of the magnetic field with the grains of magnetite reaches the central nervous system of the organisms, affecting their behavior, is still not understood.

Magnetic Resonance: Dancing Spins

We have seen that in the same way that an electron carries magnetism, atomic nuclei may also have a 'magnetic moment', albeit a much smaller one. Since this nuclear magnetism is much weaker than the normal magnetism of matter, which arises from the electrons, its effects are not easily perceived. The role of nuclei in the magnetism of matter is so small that only refined experimental techniques developed in the mid-20th century allowed the detection and analysis of this contribution. The phenomenon that most easily demonstrates the presence of nuclear magnetism is nuclear magnetic resonance (NMR), which constitutes the physical basis of the magnetic resonance imaging technique (MRI).

Atomic nuclei in the presence of an applied magnetic field circle around the field direction like a top, with a frequency of rotation, or precession, that is characteristic of the type of nucleus, and is also proportional to the intensity of the magnetic field. If a radio wave of the same frequency is made to impinge on an ensemble of precessing nuclei, it will be absorbed. This phenomenon is resonant, which means that the absorption occurs only if the frequency of the wave is the same as the turning frequency of the nuclei. Therefore, by varying the frequency of the wave and recording the variation of its absorption rate, an experimenter can determine with great precision the frequency of precession.

In the summer of 1940, a British mission led by Sir Henry Tizard (1885–1959), of the British Government Scientific Advisory Committee, visited the USA to discuss the forms of pooling the scientific resources of the two countries against the German forces, in the early days of World War II. They took with them a new device for the production of radio waves that had been developed in the previous year at the University of Birmingham, the 'magnetron'. In the magnetron, a permanent magnet made the electrons move in circular orbits, in which they emitted electromagnetic waves (microwaves) of much higher power than the previously employed radio sources. The magnetron was vital for the development of radar, a technique that helped the allies win the war. For this reason the magnetron was hailed in America as "the most valuable cargo ever brought to our shores"[22].

One of the practical peacetime outcomes of the invention of the magnetron and the technical developments associated with the technology of radar was the design of the microwave oven; this home appliance is the most common use of the magnetron. Another important consequence was the discovery of NMR*.

The first attempts to observe NMR were made[23] by the Dutch physicist Cornelius Jacobus Gorter (1907–1980) in Leiden, in 1932. He then tried to detect the resonance through a rise in temperature of the samples on absorption of the radio waves, but the sensitivity of the experiment was not sufficient to allow a positive result. A second experiment, carried out in 1942 with L. J. F. Broer using a different technique, again failed, paradoxically from an experimental difficulty associated with the fact that the samples used were too pure.

In 1938, the American physicist Isidor Isaac Rabi (1898–1988) of Columbia University published with collaborators an account of the nuclear magnetic resonance effect using a beam of molecules of lithium chloride moving in vacuum[24]. The first investigator ever to report the experimental observation of magnetic resonance in a condensed phase, in this case not the resonance of nuclei, but of electrons in a chromium salt, was Evgeny K. Zavoisky (1907–1976), of the Kazan State University, in Russia, in 1944[25].

The positive observation of nuclear magnetic resonance in solids or liquids was only reported in 1946, by two American groups working independently[26]: Edward Mills Purcell (1912–1997), Henry C. Torrey and Robert V. Pound (1919–), at MIT, and Felix Bloch (1905–1983), William W. Hansen (1909–1949) and Martin Packard, at Stanford University.

Nuclear magnetic resonance (NMR) has been applied for several decades as a powerful analytic tool in chemical studies, this use arising essentially from the fact that the nucleus of the atoms resonates or precesses at slightly different frequencies, depending on its chemical environment. Although most nuclei of several elements have magnetic moments and therefore can be used in nuclear resonance experiments, most NMR chemical studies are made with hydrogen nuclei, and a smaller proportion with the carbon isotope of mass 13. The NMR spectra of molecules give information on the number and location of the hydrogen atoms. Modern so-called Solid State NMR has been applied to studies of the dynamics, structure and morphology of new materials, such as polymers, proteins, glasses, ceramics and liquid crystals.

In magnetically ordered solids, e. g. in magnets, the NMR technique has been applied in structural studies exploiting the influence on the

* The phenomenon of magnetic resonance was first suggested by the Russian physicist J. Dorfman, in 1923 (J. Dorfman, *Einige Bemerkungen zur Kenntnis des Mechanismus Magnetischer Ercheinungen*, Zeitschrift für Physik, vol. 17, 1923, pp. 98–111.)

magnetic resonance of impurities, structural defects, and changes in magnetic order. The NMR spectrum therefore allows the probing of the immediate neighborhood of the resonant nucleus. In these ordered solids, the magnetic moments of the electrons generate large magnetic fields at the nucleus, and thus NMR is observed without the need of applying external magnetic fields.

In the 1970s, the power of NMR as a technique for the production of images was developed, beginning with the work of the American chemist Paul C. Lauterbur (1928–), then at the State University of New York[27], who reported the image of the cross-section of two vials containing water. This result, after many advances made in several different research laboratories, notably by Peter Mansfield (1933–) at Nottingham University, in England, led to this remarkable diagnosis tool, now known as magnetic resonance imaging, or MRI. MRI is based on the fact that in principle the intensity of the nuclear resonance signal is proportional to the number of resonating nuclei present in each region of a sample. By applying a magnetic field that varies from point to point, one obtains signals that arise from each small volume element of the sample. By showing on a screen the intensity of the resonance signal at each of these volumes within the sample, one can produce an NMR image. Different images are obtained for each plane of interest that intersects the sample.

In the most common use of MRI, the resonating nuclei are protons, which are the nuclei of hydrogen atoms; the intensity of the signal from each region therefore measures the number of hydrogen nuclei. In the medical applications of MRI, the technique is applied to the human body and the intensity of the signal shows the percentage of water (where most of the hydrogen atoms are located) in the soft human tissues, in the form of a computer image.

The magnetic resonance image is also sensitive to time evolution information. For example, after the application of the short pulses of radio waves, the time it takes for the resonance signals to die out can also be recorded, and this may vary from one kind of tissue to the other, allowing a differentiation, for example, of the tissues of the brain: in the brain, these decay times are different for white matter and gray matter, and for the cerebrospinal fluid[28]. The evolution of the magnetic resonance signal is also sensitive to motion of the resonant nuclei, and thus the technique can be used to detect, for example, blood flow.

MRI is less invasive than X-ray computer tomography, or CT, since the radio waves do not have the serious health hazards associated with the X-rays. The requirement of an intense magnetic field, however, makes its use dangerous for patients with implanted pacemakers and metallic prostheses. To this day, after over a hundred million MRI diagnostics have been made, there is no evidence of harmful effects of the highest magnetic fields used, in the range 3–4 T[29].

The possibility of recording images of human organs can be combined with the ability of furnishing chemical information of the type provided by conventional NMR, in a variation of the technique called Functional MRI (fMRI). This allows the visualization of the activity of different tissues, for example, of the brain, through changes in chemical composition or in blood flow. For example, after injecting into the blood stream a substance containing gadolinium atoms (the 'contrast medium'), the gadolinium magnetic moments affect the resonance frequency of the protons, and consequently the protons in the neighborhood will be 'out of tune'. This effect will make the tissue where it occurs stand out in the magnetic resonance image. This emphasizes the areas where the blood flow has increased. In the case of the brain, changes in neural activity are associated with changes in blood flow, and therefore this effect provides a mechanism for mapping the brain areas that are more active in a given situation. Also, even without the introduction of contrast media, regions irrigated with blood with different degrees of oxygenation will differ in brightness, since the magnetic properties of the hemoglobin molecules (responsible for the transport of oxygen in the blood) are modified by the oxygenation[30].

Another promising field of investigation using NMR is quantum information and computation[31]. The possibility offered by the nuclear resonance technique of changing the orientation of nuclear magnetic moments in a magnetic field has suggested that NMR can be used in computing. An ensemble of nuclei with 'spin up' or 'spin down', can represent a binary number, and since NMR pulses can change their orientation, therefore changing the stored bits, they may perform mathematical operations on these data.

A digital computer performs calculations using electronic circuits called logic gates, which are able to apply elementary operations to the binary data. For example, one such gate gives as output a '0' if there is an input of '1', and vice-versa: this is the NOT gate. Another gate gives a '1', or yes, if there is a '1' in any of two inputs – this is the OR gate.

The inputs in the digital computers are electric voltages: an electric voltage 'on' corresponds to the digit '1', and '0' otherwise. Combining different types of logic gates, every arithmetic and logic operation can be performed.

A quantum NMR computer, in turn, also requires several different gates, which in its case are not electronic circuits that respond to values of electric voltage, but rather combinations of pulses of radio waves and certain states of nuclear spins. For example, a NOT gate is made by applying a certain sequence of pulses to an ensemble of nuclear spins that had been put in a pre-established state. A great advantage is that the quantum character of the individual nuclei naturally allows the coherent superposition of states (Chapter 5), and therefore the quantum bits, or 'qubits', can store more information than ordinary bits. Quantum mathematical procedures, or algorithms, benefit from quantum laws, allowing the simultaneous processing of a superposition of states, in a naturally parallel processing mode. Using the example given in Chapter 5, this is equivalent to tossing a coin and examining with one single experiment its two sides, useful in distinguishing a true coin from a false one (one with two heads, for example).

Another application was demonstrated in 1998, with a two-qubit quantum computer employing the NMR of a solution of molecules of chloroform ($CHCl_3$). This computer successfully conducted a search in a string of objects, finding the desired object – in this example, a nucleus in a prescribed state. This type of search can in principle be completed in a very short time, thanks to the properties of the quanta.

The NMR technique has also been used in a further application of quantum computing, the division of a number into a product of prime numbers[32]. This factorization as well as other mathematical operations can be performed much faster than by classical computing modes, since in principle the power of a quantum computer would increase as the number two to the power of the number of qubits, whereas the performance of a classical computer is simply proportional to the number of bits. This has very important practical consequences, since the security of data transmission depends on encryption of data, which is nowadays mostly done using the RSA (Rivest-Shamir-Adleman) code, based on the difficulty of factoring large numbers into primes. Factoring large numbers is an extremely time-consuming task, when performed by the usual means: the fastest computer in existence today would need billions of years to factorize a 400-digit long number, a task a quantum computer would complete in about a year[33]. The promise of the quantum computer is so great that the practical realization of such a computer with this power of factorizing large numbers would immediately result in the total vulnerability of every known cryptographic system presently in use.

This chapter opened with the discussion of the remarkable assertion that the Earth and many other celestial bodies are huge magnets. As a proof of the extraordinary range of magnetic phenomena observed when the facts of nature are studied in widely different scales, the chapter closes with the dream of quantum computers, machines that some day in the future may be built based on the magnetism of atomic nuclei.

Further Reading

R.T. Merrill, M.W. McElhinny and P.L. McFadden, *The Magnetic Field of the Earth*, Academic Press, San Diego, 1998.

E.V. Mielczarek and S.B. McGrayne, *Iron, Nature's Universal Element*, Rutgers University Press, New Brunswick, 2000.

R. Wiltschko and W. Wiltschko, *Magnetic Orientation in Animals*, Springer-Verlag, Berlin, 1995.

M.E. Smith and J.H. Strange, *NMR Techniques in Materials Physics*, Measurement Science and Technology, vol. 7, 1996, pp. 449–475 (more technical).

Chapter 7
Supermagnets

"Despite our training, we rarely pause to reflect upon the reality that magnetism applications permeate our whole modern society *in its basics of electric power, communications, and information storage."*

I. S. Jacobs, in *Role of Magnetism in Technology* (1969)[1].

Introduction

The development of alloys and compounds with favorable magnetic properties, initiated in the 19th century and finally reaching a firm scientific basis in the 20th century, is a modern example of the age-old quest for useful materials. From prehistory, this relationship of the human species with matter in these specific functional forms began with the use of naturally-occurring materials, and increasingly shifted to the variety of materials designed by man. The practical use and the processing of materials have ever since been a measure of material progress.

As mankind began to try different materials to create useful objects, and also followed practical routines that guaranteed survival of the species, technology was born. This occurred long before the dawn of science, characterized in its earlier forms by the systematization of knowledge, and in later developments by the formulation of hypotheses, followed by their active testing.

In 1836 the Danish archeologist Christian Jürgensen Thomsen (1788–1865) proposed the Three Ages system of classification of the major periods of human civilization: Stone Age, Bronze Age and Iron Age. Although the time of occurrence of these periods varies from region to region, this system was universally adopted and reflects the fundamental importance of materials, especially in the fabrication of tools and implements. Of these great periods, the Stone Age was the

From Lodestone to Supermagnets. Alberto P. Guimarães

ISBN: 3-527-40557-7

longest, lasting some two and a half million years. The Iron Age started about three and a half thousand years ago, and it continues to the present time, although one often refers to a Silicon Age, called after the role of silicon-chip-based electronics in modern life.

Materials and the Conquest of Nature

The early hominids probably employed branches and stones as tools, as some modern primates do; they were initially used as needed, and afterwards discarded. It is likely that after some time pieces of wood or stone were used more permanently as implements for digging roots, or weapons for hunting animals. The first man-made tools that have survived to modern times, from about two million years ago[2], are the stone-age hand-axes, used for cutting, scraping and carving, made by chipping off flakes from pieces of flint or other hard fine-grained stone. Stone was therefore the earliest natural material of which we have evidence of adaptation by humans for practical purposes.

In the Old Stone Age, or Paleolithic period, cave dwellers discovered the use of pigments, and employed them, especially iron oxide mixed with fat or water, to create the impressive paintings that are preserved in the caves in Spain, France, and parts of Africa. These images, representing different animals, some of them extinct, and more rarely people, were made some 35 000–10 000 years ago.

With the advent of the New Stone Age, or Neolithic, beginning in the ninth-seventh millennium BC, rapid changes in the life of the communities, with the introduction of agriculture and domestication of animals, called for the use of new materials for the construction of more permanent dwellings, and for implements employed in the agricultural activities. With the beginnings of farming, containers were necessary to store grain and for cooking, and this stimulated the production of the early unfired clay vessels[3]. With the end of nomadic life, the change in lifestyle expanded the use of pottery, once its fragility became a less relevant consideration[4]. Pottery was fired on open bonfires or in kilns[5]; it was gradually perfected, probably by experimentation. The firing of clay is remarkable as the first case of an intentional operation by humans that changes the physical properties of an inorganic material*. The high temperature used for firing pottery induces a partial transformation into glass, or vitrification, that binds the grains of clay and leads to rigidity and impermeability.

Pottery is also important since its decoration gave rise to the introduction of the technique of glazing, which probably led to the discovery of glass, a novel and important material. The search for new colors possibly opened the way, through trial and error with different minerals, to the discovery of the extraction of metals from the ores, sometime at the end of the fifth millennium BC. The practice of firing pottery in kilns provided the early technical knowledge that was required for smelting (or melting a metal from the corresponding ore), which would subsequently allow the development of metallurgy.

The first metals used by humankind were naturally occurring, or native, silver and gold, later on native copper and meteoritic iron. Copper was the most abundant native metal[6]. Iron was initially extracted from meteorites, a rare source of the metal, where it is usually found alloyed with a small proportion of nickel. The relevance of this early supply of iron is evident from the name given to the metal[7] by the Sumerians – 'heaven-metal', or 'black copper from heaven', by the Egyptians.

Copper is more commonly found in nature as an oxide (copper combined with oxygen) or a sulfide (copper in a compound with sulfur). Copper oxide is easily turned into metallic copper (in a process called 'reduction') by heating to 1100 °C, in the presence of charcoal or in the absence of air. The metal obtained from the ores was used to make tools from about 4000 BC, replacing those made of stone. Copper takes its name from *cuprum*, or "metal of Cyprus", a denomination given by the Romans. Some minerals used as sources of copper, like enargite, have copper and arsenic combined, and consequently many copper artifacts from the fourth to the third millennia BC are in fact made from a copper-arsenic alloy, which is superior to the pure metal in its mechanical properties in the wrought form[8].

* According to Kranzberg and Smith (Melvin Kranzberg and Cyril S. Smith, *Materials in History and Society*, Materials Science and Engineering, 37 (1979) 1–39), perhaps the first human activity designed to transform materials to human needs was cooking; cooking also led to the very early development of materials for pans and pots. In this Section we have used this reference extensively, as well as S. L. Sass, *The Substance of Civilization: Materials and Human History from the Stone Age to the Age of Silicon*, Arcade Publishing, New York, 1998.

Copper (symbol Cu) alloyed with tin, with 90–95% Cu/weight, constitutes bronze, an alloy that with time gradually replaced Cu and Cu-As (copper-arsenic) as the raw material for tools and weapons. Bronze has the advantage of a melting point lower than that of pure copper; another (unintentional) benefit in its use is that working with bronze the early metallurgists were freed from the risks arising from the toxicity of arsenic.

The period known as the Bronze Age began about 3200 BC. This Age is considered to last until about 1200 BC, when in the Mediterranean area iron objects become more common and the Iron Age opened. However, the use of bronze objects declined gradually, without a sharply defined end[9]. An important factor that favored the substitution of bronze by iron was the widespread occurrence of iron ores, mostly in the form of oxides or sulfides[10]. The metallurgy of iron was introduced by the Hittites, in Anatolia[11], in the middle of the second millennium BC. The first iron metallurgists did not produce molten iron, since iron melts above 1500 °C, at temperatures higher than those attained in the primitive smelters. Instead, a spongy material resulted from the heating, and containing as impurities iron oxide and iron silicate; the spurious phases were removed by hammering and reheating, leading to purer iron.

Iron played an important role in the early history of the Mediterranean peoples, since the manufacture of iron weapons gave a distinct military advantage. For example, according to the Bible, under the reign of Saul, the first king of Israel, in the 11th century BC the Hebrews did not have iron swords, but the Philistines had already mastered this technology[12]. The Philistines, another people that disputed the control of Palestine, had probably learned iron metallurgy in Anatolia. Their ensuing technical superiority is an important factor in explaining why the Philistines finally defeated and killed Saul and his sons in 1000 BC.

Although iron was at first inferior to bronze in its mechanical properties, iron containing carbon, or carburized iron, is much stronger than bronze. The early smiths learned empirically that heating the metal next to pieces of white-hot charcoal improved its strength in this way. However, the understanding of this relationship between the carbon content and the physical properties of the alloy was not established until much later, in the modern era. Another technical development by the early metallurgists was the finding that rapid cooling of the iron pieces with water, a process known as quenching, gives rise to a harder material. This was already known in the 7th or 8th century BC, since for example in the Odyssey, the description of the death of the one-eyed giant Polyphemus, killed with a burning olive trunk thrust into his eye, reads[13]: "As a blacksmith plunges an axe or hatchet into cold water to temper it – for it is this that gives strength to the iron – and it makes a great hiss as he does so, even thus did the Cyclops' eye hiss round the beam of olive wood, and his hideous yells made the cave ring again."

During the Stone Age, building materials included thatch, tree branches and animal hides. Later, mud-bricks were extensively employed in the Middle East, the large-scale use of masonry in the construction of temples and monuments beginning in Egypt in the second half of the third millennium BC. In Greece and Rome the use of stone as a noble building material was further developed. Romans contributed to the progress of materials with the invention of hydraulic cement, a waterproof binding material that was employed in buildings, aqueducts and bridges. It was made from a mixture of lime and volcanic ash that hardens with the addition of water. Romans also invented concrete, by mixing cement with stones. Their great engineering works, some of which remain standing to this day, for example the Coliseum and the Pantheon in Rome, were made possible by the use of concrete.

The first alchemists, such as Zosimos of Panopolis and Synesius, active in Egypt in the 3rd century AD, described several chemical apparatuses and chemical practices[14]. In their search for the transmutation of base metals into gold, the alchemists expanded the frontiers of the knowledge of materials and perfected many chemical techniques; this is particularly true for the Arabic school, which flourished in the IX and X centuries[15].

In the Middle Ages, the accumulated knowledge of facts on materials was summarized in the *Diversarum Artium Schedula* (*"Treatise on Divers Arts"*) (1123), by a German Benedictine monk and artisan who wrote under the pseudonym of Theophilus. It contained the first European description of the techniques of casting bronze bells[16]. In the 16th century the book *De la Pirotechnia* ("Concerning Pyrotechnics") was published in Venice in 1540, by the Italian metallurgist Vannocio Biringuccio (1480-c.1539), dealing with metallurgical techniques, glassmaking and the manufacture of gunpowder. The German scholar

Georg Bauer (1494–1555), also known by the Latin name of Georgius Agricola, published *De Re Metallica* ("On Metals") in 1556. *De Re Metallica* contained a wealth of information on minerals, mining and smelting, representing an important advance in the approach to the subject of mineralogy.

Although significant technical advances were made in the use and processing of materials in the following centuries, the scientific development that marks the seventeenth-century scientific revolution had practically no impact on the understanding and improvement of the materials in use. This is probably because the complexity of the scientific problems that still had to be solved at the time concerning the structure of matter in general, and in particular the structure of complex materials, precluded any significant progress in practical applications.

In the 18th century, the discovery of the importance of the presence of a small amount of carbon in iron for forming steel opened the way to the fine control of the mechanical properties of this fundamental material. From the mid-18th century, with the investigation of the first images of steel under the microscope that inaugurated the technique of metallography, the importance of structure in the study of materials was recognized, side by side with the relevance of chemical composition. In general, during the eighteenth-century industrial revolution that began in England, progress in science and in technology followed parallel lines, usually with little mutual influence.

From the end of the 19th century, new developments in metallurgy allowed the preparation of different iron alloys that came to satisfy a growing demand from the new industries, in particular the infant car industry. At the same time, advances in organic chemistry created the first artificial dyes; further progress in the first decade of the 20th century led to the discovery of the first plastics (1909), and ten years later, the first synthetic fibers. Additional advances in synthetic materials resulted in an explosion of new products that completely changed the lives of 20th century consumers. Since plastics are formed of large molecules, they are usually referred to as polymers, from the Greek, "having many parts". Polymers today, together with metals, semiconductors and ceramics, constitute one of the major classes of useful materials.

The technique of study of the structure of matter employing X-rays, introduced in the first decades of the 20th century, allowed enormous strides in the knowledge of solids, and as a consequence, a better understanding of materials with practical applications. The structure of materials on the atomic scale could in this way be correlated with their physical properties.

The Neolithic period marks a turning point of the relation of early humans with nature, since the raising of animals and crops represents a rupture with the more passive relationship previously displayed towards the natural world by the hunter-gatherer populations. If one takes an overall look at the history of technology, it becomes apparent that one of the main avenues of technical progress, ever since the Neolithic, has always been the process of learning how to use materials, at first the naturally-occurring ones, shaped by muscular force. With time, new materials were created, and energy was tapped from other sources, such as water, wind, and domesticated animals. In every instance of technological development, the early inhabitants of the planet used natural materials such as stone, iron ores, and natural forces such as the force of wind or water, or the energy released by burning wood. Controlling, or more appropriately, conquering these natural forces was the path followed by humans towards the development of the material basis of civilization.

The metaphor of conquest, of a battle in which Nature resists in relinquishing its riches to the use of civilization, has often been employed in the description of the ways of overcoming the great obstacles to human progress. This idea, often associated with the role of technologies in the quest for utopia, was first developed during the Renaissance[17], and found its full expression in the 17th century, in the works of René Descartes (1596–1650) and Francis Bacon (1561–1626).

"Our main objective", writes[18] Bacon in the *Novum Organum* (1620), "is to make nature serve the business and conveniences of man". In the Preface to *The Great Instauration* (1623) he argues[19] that nature is "... to be conquered but by submission ..." In the same vein, according[20] to René Descartes, one should look for "a practical philosophy by means of which, knowing the force and the action of fire, water, air, the stars, heavens and all the bodies that environ us, as distinctly as we know the different crafts of our artisans, we can in the same way employ them in all those uses to which they are adopted, and thus render ourselves the masters and possessors of nature".

In that period the idea of a Nature that has to be conquered, replaces the vision of a generous Mother Nature, in whose entrails metals are generated that have to be delivered by miners. Bacon considers that Nature's products and secrets have to be obtained by force and torture, and some of his strong metaphors are reminiscent of practices of interrogation in the trials of witches[21].

Three hundred years after Bacon, the philosopher and revolutionary Karl Marx (1818–1883) again emphasized the importance of the domination of nature, for him an act that allowed man to attain the full human condition. In '*Capital*', he argues that[22] man "opposes himself to Nature as one of her own forces". Furthermore, "By thus acting on the external world and changing it, he at the same time changes his own nature." Another thinker of the 19th century, the British philosopher John Stuart Mill (1806–1873), writing in *Three Essays on Religion* (1874), expressed[23] a similar idea, in a somewhat weaker form, stating that "the very aim and object of action is to alter and improve Nature."

The mastering of the properties of matter and of techniques of preparation were also necessary requirements in the specific case of the development of modern magnetic materials. In the design of materials with desirable magnetic properties, the workings of nature have been put to practical use in such a way that progress has occurred on a scale only found in a handful of other areas of technical endeavor.

Magnetic Characterization of Materials

The overall magnetic properties of materials are usually determined from the curve of magnetization measured as a function of applied magnetic field. This curve usually has the shape given in Figure 7.1. The curve is drawn from the point 'O', which corresponds to zero magnetization for zero applied field. As the field increases to its maximum value H_{max}, the magnetization evolves along the line OA – the curve OA is called the virgin magnetization curve.

As the applied magnetic field is decreased from the maximum value, the magnetization follows the line AB; as the applied field increases in the negative direction (reaching the value $-H_{max}$) and back to H_{max}, a closed curve is drawn. Since the typical behavior of the magnetization can be described as 'lagging behind' the value of the magnetic field, the full curve is called hysteresis curve, or hysteresis loop, from the Greek *hysterein*, to be behind.

The main parameters that can be derived from the hysteresis loop are the saturation magnetization (M_s), i.e. the value of M for which the magnetization saturates, (at the field H_{max} in the figure), the remanence M_r, which is the value of M for $H = 0$, and the coercivity H_c, the value of the magnetic field for which the magneti-

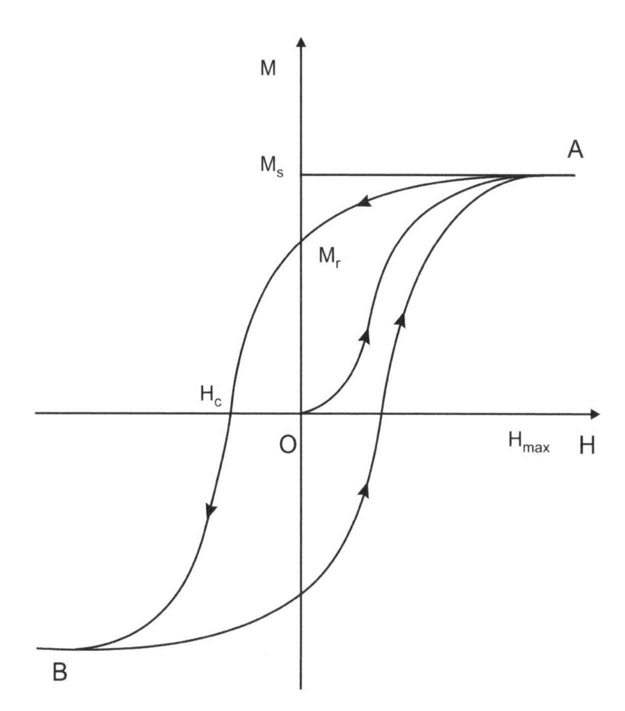

Figure 7.1 Hysteresis curve of a ferromagnet; it shows the value of the magnetization as a function of applied magnetic field. H_c is the coercive field, and M_s is the saturation magnetization.

zation M is zero. If one starts from a fully magnetized sample ($M = M_s$), the coercivity measures the magnetic field that has to be applied to completely demagnetize it.

A good permanent magnet material, or hard magnetic material, must have the largest possible values of M_s, M_r and H_c. One single parameter that sums up these characteristics is the area of the maximum rectangle that can be inscribed in the second quadrant of the curve (the region with M positive and H negative). This is known as the maximum energy product.

A soft magnetic material, on the other hand, is one that shows the least possible residual magnetization when the applied magnetic field is removed; it should have minimum values of M_r and i_c. The area of the loop is then minimal. As one varies the field, making the sample pass through a full hysteresis loop, the magnetic energy is dissipated as thermal energy. Since this energy lost as heat is proportional to the area of the loop, for soft materials the energy loss is also minimized. Soft magnetic materials are mostly used as transformer cores.

Some of the magnetic properties of materials, for example, the temperature above which there is no magnetic order (the Curie point), and the maximum value of magnetization, depend mostly on the composition of the material, and are called 'intrinsic'. Other properties depend essentially on how the material is structured on a microscopic scale, for example, on the size and shape of the small crystals that constitute the bulk; these are the 'extrinsic' properties, and include the coercivity H_c and the energy product. Typical values are shown in Table 7.1.

Material	Typical Coercivity ($kA\ m^{-1}$)	Typical Energy product ($kJ\ m^{-3}$)
Lodestone	1–10	0.01–0.1
Carbon steels	4	2
Cobalt steels	20	8
Alnico	120	40
Hard ferrite	300	36
$SmCo_5$	4000	200
NdFeB	1000	400

The Development of Magnetic Materials: the End of the Cookbook Days

The earliest known magnet, naturally occurring lodestone, is mostly composed of the iron oxide magnetite, of chemical formula Fe_3O_4. It usually contains[24] other magnetic iron oxides as well, such as maghemite (γ-Fe_2O_3). Magnetite is a relatively abundant mineral, of dark color, that forms crystals in the shape of octahedra, or octahedra modified by dodecahedral faces. Other chemical compounds of iron appear in many different minerals, iron being very abundant in the Earth's crust, as the fourth most common element, constituting some 5% of the crust.

The lodestone was very valuable for its importance in navigation, being either directly used as a compass, or, more often, employed to magnetize steel compass needles. Ships carried lodestones to re-magnetize the compasses whenever necessary[25]. In the 17th and 18th centuries, only very rich people could own a good lodestone, then worth a small fortune. For example, in 1711 the Dutch instrument maker Nicolaas Hartsoeker (1656–1725)[26] was asked, for a lodestone of the size of a fist, a price in guilders corresponding to a present-day amount of about US$50 000.

The quality of the lodestone was characterized by early authors from the weight of iron it could lift, the best stones holding[27] up to 0.57 kg (20 ounces) of iron. Another way of characterizing the lodestones, as well as other magnetic materials, is based on the magnitude of their coercivity (H_c) (see Table 7.1).

The magnetization of naturally magnetic magnetite may result from the Earth's magnetic field acting on molten or sedimentary rock, therefore aligning the atomic moments and producing a remanent magnetization (Chapter 3), or it may arise from the powerful electric currents that flow through the ground when lightning strikes[28].

From the end of the 16th century, lodestones were made more effective by wrapping or binding them with pieces of iron, a technique that increased their lifting power by as much as a factor of three[29]. In these 'armed magnets', described by Gilbert in *De Magnete*, the pieces of iron act by guiding the magnetic field lines, therefore 'concentrating' the magnetic effects.

As we have seen in Chapter 1, very early, in the 7th or 8th century, the possibility of magnetizing pieces of iron by touching or stroking with a lodestone was used in China to make compass needles. In the 13th century, Peter Peregrinus also described this process for magnetizing the needles.

Since the earliest days when the first uses were found for natural magnetite – the lodestone – there has been an enormous evolution in magnetic materials, 'magnetic materials' meaning materials presenting magnetic order, i.e. regularly arranged magnetic moments, for example, as ferromagnets (e.g. iron), or ferrimagnets, as magnetite.

From the point of view of applications, there are three great classes of magnetic materials. The first group is formed of the materials that are used as magnets, or as they are commonly called, 'permanent magnets'. In this group, the traditional examples are magnetite and iron alloys. In the 18th century these alloys were mechanically hard, so the materials for this application are still called 'hard magnetic materials'. The second class of materials, whose importance grew from the development of the uses of electricity in the 19th century, is the group of alloys that are suitable for the construction of electrical transformers, as transformer cores. These include the electrical-grade steels, which belong to a class of materials known as 'soft magnetic materials'. The third class is formed of materials of intermediate 'magnetic hardness', such as some iron oxides, that have been used, since the mid-20th century, as magnetic recording media. Hard magnetic materials are difficult to magnetize or demagnetize; that is, they require the application of large external fields to change their state of magnetization. The opposite is true of the soft materials.

The systematic construction of 'artificial' magnets was established in the 18th century, with the magnets being made of bars or strips of steel. The London instrument makers dominated the technique of manufacture of magnets in those days. The best known of them was Gowan Knight (1713–1772), who introduced cheap artificial magnets made of iron from the 1750s; one of his magnets was able to lift 28 times its own weight[30]. Knight reached great fame and became a member of the Royal Society, using this fact to support his position in the market for compasses, which he supplied to ship captains, and above all, to the Royal Navy[31].

The materials used in the manufacture of magnets in the 18th century were carbon steels; in the second half of the 19th century, tungsten steel was introduced. In the early part of the 20th century, the best magnets were made out of cobalt steels, containing 30–40% cobalt, developed by Kotaro Honda (1870–1954) in Japan.

The build-up of empirical knowledge in metallurgy, and in the later part of the 20th century, of knowledge of the quantum theory of the solid state, led to a drastic improvement in the performance of permanent magnets. This occurred hand-in-hand with the enormous advances in the methods of analysis and characterization of metals at the microscopic and atomic levels: metallography, X-ray analysis, scanning tunneling microscopy (STM), among many others. The progress of quantum mechanics coupled with increasing computing power provided material scientists with tools to predict electrical and magnetic properties of given combinations of elements forming compounds or alloys*. The science of materials was born, representing a new standard in this field, to some extent overcoming and breaking with the previous empirical tradition. In relation to the research on metallic alloys, this marked the end of what has been called by a physicist[32] "cookery book metallurgy".

In the 1930s, Tokushichi Mishima (1893–1975) of Tokyo University experimented with ternary alloys containing nickel and iron and thus discovered new materials adequate for the manufacture of permanent magnets[33]. A typical composition of these new alloys was 10 at% (atomic percent) aluminum, 25 at% nickel, 65 at% iron. With the addition of cobalt and copper, this marked the birth of the Alnico family of alloys. The values of the magnetic parameters that characterize these alloys had an important increase. This represented considerable progress, since it allowed the fabrication of smaller and more potent magnets. Further improvement was obtained by cooling the Alnico alloy in the presence of an external magnetic field, a procedure known as magnetic annealing[34].

The magnetic properties of certain substances that are mixed oxides of iron with other metals, called ferrites, were studied since 1936 at the Philips Research Laboratories in the Netherlands[35]. This investigation continued after World War II, and the exploration of the potential for application of ferrites in permanent magnets followed suit. Ferrites were used as magnetic memory cores in one of the first digital computers, Whirlwind I, built at MIT in 1953[36].

The more common ferrites have the formula MFe_2O_4, where M is a metal; when M is iron (Fe), we have magnetite, which is therefore the earliest known natural ferrite. The ferrites substituted previous magnetic materials to a large extent, and are attractive for their low cost and high electric resistance. Ferrites can be either "hard magnets", suitable for the fabrication of permanent magnets, or "soft magnets". The "soft" ferrites require only a small field to be magnetized and are widely used in antennas, inductors, and radio-frequency devices.

Ferrites used in the fabrication of magnets are of the hexagonal variety, typically with formula $Fe_9M_{12}O$, where M is either barium or strontium. Every year millions of tons of ferrites are produced worldwide[37].

* A materials scientist characterized the relationship between the different approaches in dealing with materials in a succinct way: "condensed-matter physics (explanation), solid-state chemistry (exploration) and materials science (exploitation)." (Paul Calvert, 'Advanced Materials', in *The New Chemistry*, Ed. Nina Hall, Cambridge University Press, Cambridge, 2000, p. 352.)

Other permanent magnet materials are the alloys[38] copper-iron-nickel, copper-nickel-cobalt, platinum-cobalt, iron-cobalt-molybdenum, and also[39] manganese-aluminum-carbon and iron-chromium-cobalt. Although there are today some thousands of magnetic materials used for several different applications, a dozen of these materials are responsible for 99% of the world market[40].

The design of these new materials was combined with new techniques of preparation, where the structure of the materials at the microscopic level, or microstructure, is controlled by grinding, alloying, heat treatment, sintering (aggregating a powder under pressure and high temperature), and so on. This means not only that the chemical structure can be controlled, but also that the shape, size, and distribution of the crystallites of the compounds present in the magnet can be tailored to produce the desired properties. In addition, the degree and direction of magnetization are manipulated through the application of intense magnetic fields in the process of preparation of the permanent magnets.

The evolution of permanent magnet materials can be demonstrated by Table 7.1, where the parameter energy product is shown to have increased by factor of about 200 since the first artifical magnets were made; as this product measures the amount of energy that can be stored in a magnet, it therefore gives a quantitative measure of the quality of a magnet.

The soft magnetic materials are the materials of choice for cores in electric transformers. The first transformer was built by Michael Faraday when he wound two coils around an iron ring (Chapter 3). As an alternating (AC) voltage is applied to one coil, another AC voltage appears on the other coil, and these voltages are in the ratio as the inverse of the ratio of the number of turns in the coils. In each cycle of the AC voltage, the magnetization of the core passes through a hysteresis loop (see Figure 7.1 in Box), and part of the electrical energy is transformed into thermal energy, appearing as heat. The predominance of AC power transmission, from the last decades of the 19th century, led to the universal use of transformers, therefore turning these considerations on the magnetic properties of core materials into issues of major economic relevance[41].

In the second half of the 19th century, many iron and nickel alloys were investigated in England from the point of view of their magnetic properties, by pioneers such as the Scottish engineer and physicist

James Alfred Ewing (1855–1935), the discoverer of the hysteresis phenomenon. In the early 1900s, mostly through the work of another investigator, Robert Abbot Hadfield (1858–1940), the alloy of iron with 3–4 % silicon was introduced for transformer cores, reducing the core thermal losses to about half of that of previously used alloys[42]. Iron silicon alloys were perfected in many ways over the years, the most important improvement being the introduction in the late 1930s of alloys in plates with planes of the individual metallic crystals oriented (called 'grain oriented')[43]. Typical coercivities of these materials are of the order of 10 A m^{-1}.

Other alloys that present good soft magnetic properties are Mumetal and Permalloy, both based on iron and nickel, and discovered in the 1920s.

An important development in the direction of soft magnetic materials was the discovery and production of amorphous alloys, or 'metallic glasses', starting in the 1960s (Chapter 5). Soft amorphous alloys have been prepared containing iron, cobalt, phosphorus, bismuth, etc.

Another class of soft iron-based magnetic alloys is formed of very small crystals, of around 1/100,000 of a millimeter, embedded in an amorphous matrix – the nanocrystalline alloys, introduced in Japan in the late 1980s. These are prepared by partial crystallization of amorphous alloys.

The soft magnetic alloys have other applications besides transformer cores: they are employed in many types of sensors and as shields to avoid stray magnetic fields. They represent about one third of the global market for magnetic materials, estimated to be US$30 billion in 1999[44].

A novel terrain for magnetic materials is the field of 'spintronics'[45], the electronics that relies not only on handling and responding to the charge of the electron, but also on its intrinsic angular momentum, or spin. Spintronic devices make use of the interaction of magnetic materials with the electron spin to manipulate separately electrons with spins in one direction and spins in another direction. For example, thin films of alternate ferromagnetic and antiferromagnetic layers conduct selectively electrons with different spin direction; this is the physical mechanism behind the giant magnetoresistance (GMR) of films, an effect that is characterized by a large variation in resistance produced by an applied magnetic field. Spintronics applications include a variety of GMR-based devices, such as read heads for magnetic hard disks, magnetic field sensors, and magnetoresistive random access memories (MRAMs). Some spintronic devices may use materials that are at the same time both semiconductors and ferromagnetic. The development of spintronics and the continued progress in miniaturization hold the promise of much smaller and faster electronic components and devices.

Supermagnets

In the 1970s, a new class of permanent magnet materials was introduced based on the rare earths, elements of the Periodic Table that in many cases show magnetic order. Originally, the denomination of 'earths' was given by the Greeks to the oxides of calcium, aluminum and magnesium. The name rare earths designates the elements found from the end of the 18th century, initially as a result of the work of the Finnish chemist Johan Gadolin (1780–1852). They were then thought to constitute a class of elements with very low natural abundance, hence the name. However, some of the rare earths are relatively abundant in the Earth's crust; the most abundant of the family, cerium, is the 29th most common element in the planet's crust.

The chemical elements that receive this denomination are the series of 14 elements of the Periodic Table beginning with lanthanum (symbol La, atomic number $Z = 57$) and continuing on to lutetium (Lu, $Z = 71$), known as the lanthanides, and, in addition, the elements scandium (Sc, $Z = 21$) and yttrium (Y, $Z = 39$). Because they show great chemical similarity, rare earths are difficult to separate from each other, and were obtained in quantity only with the development of purification techniques, after World War II.

Many rare earths are highly anisotropic, i.e. their physical properties depend strongly on the direction in the crystal along which they are measured. Anisotropy implies that the magnetization prefers to remain pointing along a given crystal direction. This property is relevant for materials employed in the manufacture of permanent magnets, because one of the requirements of a magnet is that, under the action of an external magnetic field, it will resist changing the direction and the magnitude of its magnetization. Among the most successful compounds for this purpose containing rare earths, are samarium-cobalt and neodymium-iron-boron compounds. The rare-earth-based materials now represent some 30% of the world market (in value) for permanent magnet materials[46].

An early indication of the promise of rare-earth-containing hard magnetic materials was the discovery in 1935 by a Saint Petersburg group, in Russia, of energy products above 340 kJ m^{-3} for a neodymium-iron alloy[47]. A long period elapsed until the first rare-earth-containing magnets were considered. These were based on the series of cobalt compounds of chemical formula RCo_5 (where R is a rare earth), whose magnetic properties had been studied in the 1960s[48].

The second generation of rare earth permanent magnet materials was based on the compounds of general formula R_2Co_{17}. A related series of intermetallic compounds, the compounds R_2Fe_{17}, resulted in useful materials when combined with elements with small atomic radius (hydrogen, nitrogen and carbon), whose atoms occupy the interstices of the crystal lattice.

The third and most remarkable family of rare earth magnets was developed simultaneously[49] in 1983 by General Motors in the US, and Sumitomo in Japan, and it contained neodymium, iron and boron, in a compound with chemical formula $Nd_2Fe_{14}B$. These new materials, known as high-performance magnets, or "supermagnets" (a designation also applied to all magnet materials containing rare-earths), have very large energy products (see Table 7.1). An additional advantage of the magnets with this formula is that iron is less expensive than cobalt, and neodymium is less expensive, for example, than the samarium used in $SmCo_5$.

As with all the other permanent magnet materials, complex processing is required to turn the NdFeB alloy into a usable magnet: the magnets are usually made from the oriented powder pressed into the desired form (sintered), or by rapidly cooling the alloy from the melt into flakes, that are resin- or polymer-bonded into the desired shape[50]. Nowadays, industrially produced NdFeB alloys have compositions near $Nd_{15}Fe_{77}B_8$.

Magnetic materials have undergone a spectacular evolution[51] during the 20th century. The energy product has doubled every 12 years, the maximum permeability (a measure of how easy it is to magnetize a given material) has doubled every 6 years, and in connection with the latter fact, transformer core losses were halved every 8 years. Throughout the 20th century, the maximum energy product of permanent magnetic materials increased by a factor of 200.

All these advances in materials stimulated the widespread use of permanent magnets in electric motors, sensors, transducers, and so on. The small size made possible by the new strong magnets favored the creation of new products, like the small earphones for portable radios, CD players, etc.

Magnets exist nowadays in most home appliances: in practically every electric motor, in drives of CD, DVD and VCR players, in motors that equip washing machines, electric knives, fans, air conditioners, hard disk and floppy disk drives, garbage disposal units, toys and so on. Also in magnetic latches (as in a refrigerator door), in the focusing system of a TV set, in notice boards, switches, actuators, headphones, loudspeakers and microwave ovens.

The automotive industry employs many magnets to equip cars with sensors, motors, and so on. In a modern car, there may be some 100 magnets, as part of the starter motor, air conditioner, speedometer, gauges, washer pump, antenna lift motor, fuel pump motor, door lock motor, CD drive motor, speakers, windshield wiper motor and anti-skid system.

Cool Magnets

The temperature of a piece of paramagnetic, antiferromagnetic or ferromagnetic material subjected to a magnetic field varies as the field is removed; this effect is known as the magnetocaloric effect[52]. It was discovered in pure iron, in 1881, by the German physicist Emil Warburg (1846–1931). In the most common form of the effect, the removal of the field leads to cooling. This can be explained in simple terms: when the order imposed by the magnetic field on the electronic spins is no longer present, the total order is 'conserved'; thus the thermal disorder, and with it the temperature, is reduced.

Using this effect in gadolinium sulfate, in a process known as adiabatic demagnetization, low temperatures of the order of one thousandth of a degree above absolute zero have been attained. Still lower temperatures, of one-hundred-thousandth of a degree or lower, were reached employing the analogous nuclear effect. In this case, the magnetic field imposed some magnetic order on the system of nuclear magnetic moments, not electron moments, which on demagnetization couple to the rest of the sample, cooling it.

Since 1997 some compounds that present a large magnetocaloric effect have been found (e.g. $Gd_5Si_2Ge_2$, $MnFeP_{0.45}As_{0.55}$); these compounds have an enhanced ability to remove heat. In ferromagnetic materials, the magnitude of the magnetocaloric effect is larger near the transition temperature above which there is no magnetic order (the Curie point). Therefore, the most interesting materials considered for refrigeration applications should have transition temperatures in the range of the planned operating temperatures; to substitute a home refrigerator, the transition temperature should be near room temperature.

This effect was demonstrated in practice in the operation of a magnetocaloric refrigerator, where a chunk of magnetocaloric material is moved in and out of the magnetic field, and a heat exchanger allows it to cool a specified volume. The first working prototype[53], with a cooling power of 500 watts, was built by Astronautics Corporation in the USA, in 1997. It used a helium-cooled superconducting magnet to produce the magnetic field. In 2001, a more practical refrigerator employing permanent magnets was also demonstrated.

The appeal of this novel technique is that there are no polluting or toxic refrigerating gases, and furthermore, there may be a significant reduction in the number of moving parts of the refrigerator. The efficiency that can be attained, in the range of 50–60%, is higher than that of the usual compressor refrigerator. Magnetic refrigeration seems to be a promising technique to be used for gas liquefaction.

Magnetic Recording: Magnets that Remember

In the novel cycle *Remembrance of Things Past*, the masterpiece of the French writer Marcel Proust (1871–1922), there is a famous passage[54], inspired by an episode experienced by the author. In it, the central character describes how a chain of images connected to his youth is awakened by the smell of a cake – a 'madeleine': "And as soon as I had recognised the taste of the piece of madeleine (...) immediately the old gray house upon the street, where her room [his aunt's] was, rose up like a stage (...) and with the house, the town, from morning to night and in all weathers, the Square where I used to be sent before lunch, the streets along which I used to run errands, the country roads we took when it was fine."

In the example of Proust's novel, information stored in the brain is evoked many years later by an external stimulus, the smell of a madeleine; other recollections may arise from an image, a name, or a sound. Despite the ability of the brain to store a prodigious amount of information, human memory may fail, and a secure means is required for the recording of data and the transmission of knowledge. One might say that mankind invented writing in response to this need; this great invention first appeared among the Sumerians, in Mesopotamia, in the fourth millennium BC.

The written word created a revolution in the diffusion and accumulation of knowledge, and led to the invention of the book, initially consisting of clay tablets or papyrus rolls. The library of the Museum of Alexandria, one of the most celebrated institutions of the ancient world, was founded in the 3rd century BC and destroyed over the following centuries. At the height of its existence it contained scrolls with contents corresponding to some 50 000 average books[55], a remarkably rich collection for the period.

How can one quantify the amount of information contained in a book? To specify one character of text (letter, space or punctuation mark) one needs some eight units of data, or 'bits' (from 'binary digit'), each of which can be represented by a '0' or a '1'. A set of eight bits corresponds to one 'byte'. It is estimated that an average book without illustrations contains one million bytes of information, or one megabyte.

Contrasting with the Alexandria library, a modern national library such as the US Library of Congress, has 20 million books on its shelves. They contain, in the text of the volumes alone, some 20 million million bytes, or 20 trillion bytes (or alternatively, 20 terabytes)[56].

The comparison between a modern library and the Museum of Alexandria illustrates the information explosion, which is one of the characteristic aspects of our era. This phenomenon has demanded new forms of information storage; the discovery of magnetic recording afforded a practical means of preserving and making easily available sound, text, numeric data, and images.

To illustrate the basic idea behind magnetic recording, consider lava flowing from a volcano. As the lava solidifies, the Earth's magnetic field orients those atoms that have magnetic moments, and a magnetization, called 'thermoremanent magnetization', results (Chapter 2). If now one brings close to the rock a device able to detect this magnetization, the 'written' rock magnetization can be 'read'; this is the principle behind magnetic recording.

In a magnetic recording apparatus, data are recorded when electrical impulses are translated into a pattern of tiny magnetized regions on a recording medium, for example, a film of magnetic material deposited onto a tape or a disk. This is done by first amplifying the electric signal; it then generates a magnetic field that is applied to the recording medium, creating the required magnetization. This magnetization is read by reversing the process: the magnetic field produced by the medium generates an electric signal related to the magnetization when it moves past a coil or sensor.

Recording can be either analog or digital: in the first case the magnetization created on the medium is essentially proportional to the signal that is being recorded, and in the second type the information is encoded into regions magnetized in one direction or in the opposite direction (bits '0' and '1'). The earliest examples of sound recording were of the analog type, in which the electric signal is proportional to the intensity of the sound that is being recorded.

Magnetic recording was invented at the end of the 19th century by the Danish engineer Valdemar Poulsen (1869–1942), who then worked at the Copenhagen Telephone Company (KTAS). In the Copenhagen Company, as in several others around the world at the time, there was an interest in devising a system for recording telephone messages. This stimulated Poulsen to experiment with magnetic recording, trying initially with an electromagnet that magnetized a circular saw[57]. He only succeeded in his attempts when he tried as magnetic medium a 1-meter long steel wire, held by two nails on a piece of wood. He moved along the wire an electromagnet that was connected to a microphone, magnetically recording his voice.

Poulsen perfected his idea and filed the first patent in August 1898 for a recorder that used a drum to wind the steel wire, the 'Telegraphone'. He left the telephone company shortly after and started to work on developing his projects, in association with the engineer and friend Peder Olaf Pedersen (1874–1941). The invention provoked much interest at the time, and Poulsen was awarded[58] the grand prize at the Paris Exhibition of 1900. In the same Exhibition the oldest extant voice recording was made, of the Emperor Franz Joseph of Austria (1830–1916)[59].

Besides his drum recorder, he also patented a recorder using a steel ribbon. However, neither device was very practical, because the record/play time was too short, insufficient for recording telephone messages. Since no other application of magnetic recording was envisaged at the time, the inventions were largely abandoned by Poulsen, who developed, starting in 1902, an interest in wireless transmission.

Tape recording turned out to be much more practical than steel wire recording, among other reasons due to the fact that splicing a torn tape was much easier than joining a broken steel wire. The Austrian inventor Fritz Pfleumer (1881–1945) had developed a method of coating cigarette paper with tiny bronze particles, and adapted the technique to deposit powdered magnetic materials, patenting the idea in 1928[60]. The first successful magnetic recorder using magnetic tape was the 'Magnetophon', created in Germany by the companies AEG and IG Farben, and first shown in public in 1935[61]. This equipment was widely used by the German state radio network, especially to record broadcast programs so that they could be censored before transmission[62].

A magnetic recording machine using a steel ribbon was developed in Britain beginning in 1929 by a German film producer, Ludwig Blattner (1884–1935). The 'Blattnerphone', originally devised to be part of a motion picture sound system, was employed by the British Broadcasting Corporation (BBC) to record short-wave radio programs[63].

In America, the research on magnetic recording started at Bell Labs in 1930, initially on a telephone answering machine[64]. Developments in the USA included a wire recorder designed in 1939 at the Armour Institute of Technology (now Illinois Institute of Technology). In the late 1940s, the difficulties with re-joining the broken steel wires in the recorders caused the coated paper or plastic tapes introduced by the German makers to supersede the wire technology. After World War II, American companies such as Ampex perfected Magnetophons brought from Germany, using ideas of the German designers and benefiting from the fact that the German patent rights had been annulled, as part of the treaty on war compensations[65].

Since the 1950s, magnetic recording has become available for home uses, and magnetic tape was perfected and produced in the USA, Europe and Japan. The explosion of magnetic recording equipment in the consumer market came with the introduction of the cassette (today compact cassette) system by the Dutch manufacturer Philips. The cassette was presented in 1963, and established itself as the standard of sound recording systems produced all over the world, both for home use and for the car. Recorded music cassettes appeared in 1965. Another step that promoted further popularization of the small format magnetic tape was the invention of the Walkman, the portable tape player, by Sony of Japan, in 1979.

Another front in the use of magnetic recording was the reproduction of TV images on tape. This required storing much more data than in the sound-only tapes. The first commercial video recorder (VCR) was produced by Ampex in the US, in 1956[66]. Magnetic tape video recorders have since occupied a significant niche in the consumer electronics market, lately challenged by the optical digital versatile (or video) (DVD) players.

Credit cards and automated teller machine (ATM) cards also use a magnetic strip where information on the owner's account, card number, and in some cases the personal identity number (PIN), are stored. Magnetic technology in this case competes with cards with embedded chips (Smartcards) introduced later. Magnetic strips are also found in some subway tickets, plane boarding cards and other types of cards and badges.

The great expansion in the use of digital computers since the introduction of the first personal computers in the late 1970s spurred the improvement of magnetic recording systems for the storage of data. This includes the familiar computer magnetic hard disk drives (HDD) and floppy disks, or diskettes. Small rings of ferrite had been used for the first time as magnetic memory elements in IBM computers, beginning in 1955[67]. Hard disks were introduced by IBM in 1956, with data stored at an areal density[68] of 2 kbit/in^2 (3 bit/mm^2). Computer data are stored in digital form, i. e. as a sequence of '0's and '1's, corresponding to regions magnetized in opposite directions (Figure 7.2).

Hard disks and floppy disks have an essentially two-dimensional recording medium, usually a film containing cobalt-based magnetic particles deposited onto a plastic flexible disk (the floppy) or an aluminum, plastic or glass disk, in the HDD. In the case of the HDDs, a read-write head moves on an air cushion at a very small distance (of the order of ten millionths of a millimeter) from the surface of the disk, which turns thousands of times per minute. The write head records each bit in an extremely small area, of only 5/100 of a mil-

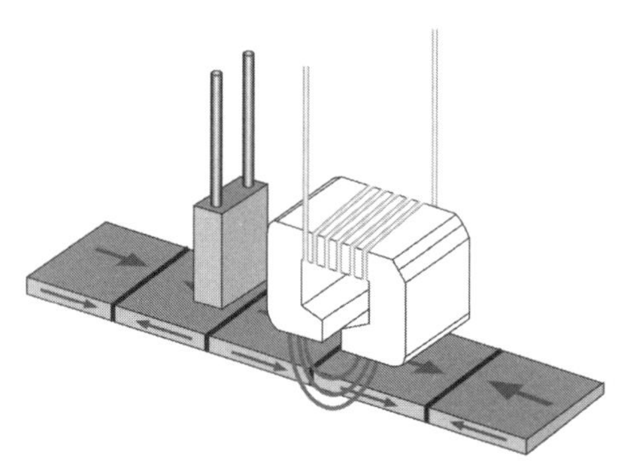

Figure 7.2 Schematic view of a reading/recording head of a magnetic hard disk. The recorded information is read through a magnetic sensor (little box on the left) that detects the magnetization state of each bit (magnetized in one way or the other, as shown by the arrows, which correspond to "0" or "1"). The recording head (right), acts by applying a magnetic field to record the disk element in the desired direction.

lionth square millimeter[69], giving a density of 10 Gb/in^2. The writing is done by generating a magnetic field across a small gap (of the order of 100 nanometer wide) in the write part of the head on the surface of the medium. In the new disk units, the 'read' head senses the magnetic field due to the magnetized disk through the change that it produces in the resistance of the material of the head. This magnetoresistance effect is the basis of the read heads introduced in the early 1990s; in 1997, heads based on a more intense effect (the giant magnetoresistance or GMR effect), with more than double the sensitivity, were introduced.

The giant magnetoresistance effect is also used as the basis of the magnetoresistive random access memory (MRAM), designed to substitute the silicon dynamic random access memory (DRAM), which is volatile, i.e. it does not retain the information when the computer is switched off. In the MRAM unit, which is non-volatile, each bit is stored as the state of magnetization of a tiny magnet, and the state ('1'or '0') is read by measuring the electrical resistance. MRAM units have very short data access times and use less energy than semiconductor random access memories, which have to be refreshed many times per second.

Magnetic recording units for computers have evolved at an amazing pace: in 40 years the gains in compressing the information into smaller and smaller space has reduced the cost per data bit by a factor of 500 000. The data recording density has doubled[70] every two years since the late 1950s! The present demonstrated areal density of data units (bits) for conventional magnetic media is of 100 thousand million bits per square inch, or 100 Gbit/in^2 (150 Mbit/mm^2), corresponding to one hundred thousand times the density attained in 1970. However, as the magnetized data unit (the bit) becomes smaller than a certain dimension, corresponding to a density[71] of the order of 100 Gbit/in^2 (150 Mbit/mm^2), its magnetism becomes unstable, i.e. the recorded data is lost after some time. Therefore, the present process of increasing information density cannot proceed indefinitely; a new paradigm has to be established.

One way of circumventing this thermal instability limitation is the use of multilayer magnetic media, with two superposed layers that are magnetized in the opposite direction (called "antiferromagnetically coupled", or AFC media). Disks with densities above 25 Gbit/in^2 using this principle have been produced by IBM[72] since 2001.

Instead of magnetizing the medium in a direction parallel to its surface, future devices may employ perpendicular recording, which in principle has the ability of attaining higher densities, but still presents technical difficulties. Other new avenues in magnetic recording will involve further innovations in the magnetic media, such as the use of patterned media, where the magnetic surface is divided into islands, made for example, through lithography, each island storing a single bit. This method may allow attaining data recording densities[73] to reach 1000 Gbit/in^2 (or 1 terabit/in^2) and beyond. Despite the fact that the atomic nature of matter sets an ultimate limitation to the continuous growth of information storage density, magnetic recording is likely to keep its dynamism for many years to come (Figure 7.3).

The rapid progress in magnetic storage technology has led to an increasing share of the information kept in magnetic form. According to a study[74] made in 2003, of the world's total estimated information output recorded under all forms (on paper, film, magnetic and optical media) – about five million terabytes – 92% was stored as magnetic data (Figure 7.4). This fantastic amount of information is comparable

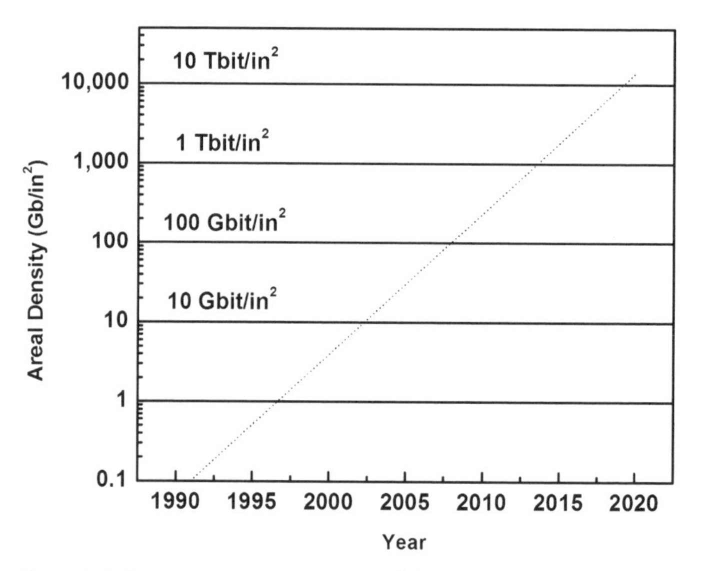

Figure 7.3 Approximate representation of the evolution of the recording density in hard disks and probable scenario for the next years. The values of the density, in Gigabits per square inch, are represented on a logarithmic scale.

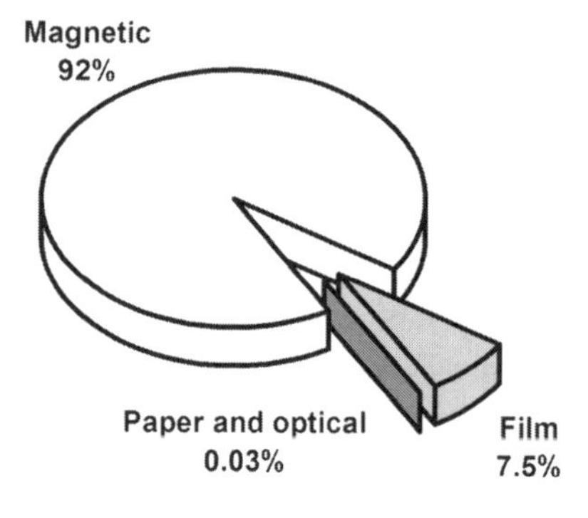

Figure 7.4 How the global recorded information in one year is distributed among the different storage media, showing the overwhelming dominance of magnetic recording.

to the information content of all the words ever spoken by people! This amount is much larger, for example, than the information contained in the approximately one million new books that are published each year. Or in the 2.5 billion CDs sold every year.

We may therefore conclude that magnetic recording technology has attained a very special position in our time, being responsible for the custody of most of the information or knowledge generated world-wide.

Conclusions

The wonder about the attraction of the lodestone has taken us a long way, from a curious phenomenon that was the object of awe and bafflement, to an aspect of nature present in every scale of complexity, from subatomic particles to galaxies. This road leads from the first practical device using magnetic forces – the compass – to wonderful tools for investigating the innermost secrets of the human body, or to novel and fantastic ways of dealing with information. From the early compass that guaranteed the safety of sailors, to the safeguard of mankind's wealth of knowledge.

I hope that I have succeeded in sharing with the reader some of the amazement surrounding the phenomenon of magnetism, which has captured the imagination of scientists and laymen alike for such a long time; and in giving, in addition, an inkling of how the advances in the knowledge of this phenomenon relate to the expanding frontiers of science.

Further Reading

Melvin Kranzberg and Cyril S. Smith, *Materials in History and Society*, Materials Science and Engineering, 37 (1979) pp. 1–39.

J. D. Livingston, *100 Years of Magnetic Memories*, Scientific American, November 1998, pp. 81–85.

J. D. Livingston, *The History of Permanent-Magnet Materials*, Journal of Metallurgy (JOM), February 1990, pp. 30–34.

S. L. Sass, *The Substance of Civilization: Materials and Human History from the Stone Age to the Age of Silicon*, Arcade Publishing, New York, 1998.

Timeline

2000 BC

- 2000–1500 BC mention of “grasping hematite” for magnetite, in a list of commodities, Mesopotamia.
- 1400–1000 BC polished magnetic bar in Olmec site, Mexico.
- 1200 BC beginning of the Iron Age in the Mediterranean region.

1000 BC

- first half of the first millenium BC: “grasping hematite” in a lexical text, Mesopotamia.
- 6th century BC Thales of Miletus (c. 640–546 BC) in Greece, according to Aristotle, considered that the magnet had a soul.

500 BC

- 5th century BC Empedocles of Acragas (d. c. 433 BC): attraction of the magnet due to ‘effluences’.
- Plato (c. 428-c. 348 BC) in Meno refers to the electric ray.
- Aristotle (384–322 BC) in his *History of Animals* describes the action of the torpedo electric ray.
- c. 400 BC Democritus (c. 460-c. 370 BC) creates atomism.
- 300 BC-AD 100 Olmec magnetic carved stone turtlehead, Mexico.
- 240 BC in China the book *Lü Shih Chhun Chhiu* (“Master Lü’s Spring and Autumn Annals”) describes the properties of the magnet.
- 1st century BC Lucretius (95–55 BC) the name magnet comes from the province of Magnesia; attraction is due to a stream of particles emitted by the magnet.

AD

- c. AD 47 Scribonius Largus (fl. 43) publishes *Compositiones medicamentorum* mentioning the therapeutic use of the electric fish.
- AD 83 the *Lun Hêng* (“Discourses Weighed in the Balance”) describes the south pointing spoon.

From Lodestone to Supermagnets. Alberto P. Guimarães

ISBN: 3-527-40557-7

- 1st century AD: Pliny the Elder (AD 23–79) in *Natural History*, wrote about the magnet.

AD 500

- 8th or 9th century: magnetic *declination* known in China.

AD 1000

- 1044 *Wu Ching Tsung Yao* ("Collection of the Most Important Military Techniques") published, with a description of the *compass*.
- 1123 Theophilus (fl. 12th century) publishes the *Treatise on Divers Arts*, with information on materials and techniques.
- 1175–1183 Alexander Neckam (1157–1217) publishes *De Nominibus Utensilium* ("On the Names of the Instruments") with reference to the *compass*.
- 1269 Pierre de Marincourt (Petrus Peregrinus) (b. c. 1220) publishes the *Epistola de Magnete* ("Letter on the Magnet") in which he summarizes the current knowledge on magnetism.

AD 1500

- 1540 Vannocio Biringuccio (1480-c. 1539) publishes the treatise *Concerning Pyrotechnics*.
- 1543 Nicolaus Copernicus (1473–1543) publishes *De revolutionibus orbium coelestium* ("On the Revolutions of the Celestial Spheres").
- 1544 Georg Hartmann (1489–1564) is the first to report the phenomenon of *inclination*.
- 1556 Georgius Agricola (1494–1555) publishes *On Metals*, a treatise on mineralogy, mining and metallurgy.
- 1558 Giovanni Battista (or Giambattista) della Porta (1535–1615) publishes *Magia Naturalis* ("Natural Magic"), with many references to magnetism.
- 1581 Robert Norman (fl. 1590) publishes *The New Attractive*, where he describes *inclination*.

AD 1600

- 1600 William Gilbert (1544–1603) publishes *De Magnete*, where he states that the Earth is a magnet.
- 1603 Accademia dei Lincei, the first scientific society in history, is founded in Rome.
- 1610 Galileo Galilei (1564–1642) publishes *Sidereus Nuncius* ("The Starry Messenger").

- 1618–1621 Johannes Kepler (1571–1630) publishes his *Epitome Astronomiae Copernicanae* ("Epitome of Copernican Astronomy"), where he compares the attraction of the Sun with that of the *magnet.*
- 1629 Niccolo Cabeo (1596–1650) publishes *Philosophia Magnetica,* with the first description of repulsion between like charges.
- 1632 Galileo Galilei (1564–1642) publishes *Dialogo Sopra i due Massimi Sistemi del Mondo* ("Dialogue on the two Chief World Systems").
- 1641 Athanasius Kircher (1601–1680) publishes *Magnes, sives De Arte Magnetica* ("The Magnet, or About the Magnetic Art").
- 1644 René Descartes (1596–1650) publishes the *Principles of Philosophy,* where he ascribes the magnetic attraction to a flow of 'threaded particles'.
- 1663 Otto von Guericke (1602–1686) builds an *electrostatic machine.*
- 1676 the compass in a ship is affected by lightning.
- 1687 Isaac Newton (1642–1727) publishes the *Principia Mathematica.*
- 17th century Valentine Greatrakes (c. 1628 – c. 1700) reports cures made with *magnets.*

AD 1700

- 1701 Edmund Halley (1656–1742) publishes the first geomagnetic chart.
- 1704 Isaac Newton (1642–1727) publishes *Opticks.*
- 1709 Francis Hauksbee the Elder (c. 1666-c. 1713) finds that low-pressure air glows with the passage of electricity.
- 1729 Stephen Gray (1666–1736) discovers that electricity can be transmitted over distances.
- 1733 Charles-François de Cisternay Du Fay (1698–1739) discovers two types of static electricity.
- 1746 Pieter van Musschenbroek (1692–1761) invents the Leyden jar.
- 1747 Abbé Nollet (1700–1770) builds one of the first electrometers.
- 1752 Benjamin Franklin (1706–1790) performs the experiment with atmospheric electricity.
- 1753 Georg-Wilhelm Richman (1711–1753) is killed in Russia while experimenting with lightning.

- 1785 Charles Augustin de Coulomb (1736–1806) reports that the magnetic force falls with the square of the distance.
- 1789 Antoine-Laurent Lavoisier (1743–1794) publishes the "Elementary Treatise of Chemistry", and mentions the existence of 33 elements.
- 1791 Luigi Galvani (1737–1798) publishes *De viribus electricitatis in motu musculari commentarius* ("Commentary on the Effect of Electricity on Muscular Motion") with his experiments with frogs.

AD 1800

- 1800 Alessandro Volta (1745–1827) publishes his discovery of the electric battery.
- 1800 William Nicholson (1753–1815) and Anthony Carlisle (1768–1840) discover *electrolysis.*
- 1820 Hans Christian Oersted (1777–1851) discovers the magnetic effect of currents
- 1820–21 André Marie Ampère (1775–1836) attributes the magnetism of matter to 'molecular' currents.
- 1831 Michael Faraday (1791–1867) discovers electromagnetic *induction.*
- 1843 Samuel Heinrich Schwabe (1789–1875) publishes evidences of the sunspot cycle.
- 1845 Michael Faraday (1791–1867) discovers *diamagnetism* and *paramagnetism.*
- 1864 James Clerk Maxwell (1831–1879) publishes "A Dynamical Theory of the Electromagnetic Field" where he describes the theory of electromagnetism.
- 1867 First Michelson-Morley experiment attempts to detect the motion of the Earth relative to the "*ether*", finding a negative result.
- 1869 Dmitri Ivanovich Mendeleev (1834–1907) publishes "The Periodic Table of Elements".
- 1881 Emil Warburg (1846–1931) discovers the magnetocaloric effect.
- 1885–1889 Heinrich Hertz (1857–1894) detects radio waves and identifies their electromagnetic character.
- 1890 James Alfred Ewing (1855–1935) discovers *hysteresis.*
- 1895 Wilhelm Conrad Roentgen (1845–1923) discovers X-rays.

- 1895 Pierre Curie (1859–1906) discovers the law of variation of the magnetism of paramagnetic materials with temperature (known as Curie Law).
- 1896 Antoine Henri Becquerel (1852–1908) discovers radioactivity.
- 1897 Joseph John Thomson (1856–1940) discovers the *electron*.
- 1898 Valdemar Poulsen (1869–1942) invents magnetic recording.

AD 1900

- 1900 Max Planck (1858–1947) introduces the idea of the quantum.
- 1905 Albert Einstein (1879–1955) publishes the special relativity theory.
- 1905–1910 Paul Langevin (1872–1946) publishes a theory of paramagnetism.
- 1907 Pierre Weiss (1865–1940) proposes the molecular field to explain ferromagnetism.
- 1913 Niels Bohr (1885–1962) publishes a model for the atom.
- 1919 Joseph Larmor (1857–1942) proposes the self-exciting dynamo theory to explain the magnetism of the Sun.
- 1922 Arthur Holly Compton (1892–1962) demonstrates the existence of photons.
- 1923–1924 Louis de Broglie (1892–1987) suggests that all matter have a wave character.
- 1925 Samuel Abraham Goudsmit (1902–1978) and George Eugene Uhlenbeck (1900–1988) propose the existence of *spin*.
- 1926 Werner Heisenberg (1901–1976) and Paul Dirac (1902–1984) propose the exchange interaction to explain magnetic order.
- 1926 Erwin Schrödinger (1887–1961) introduces his wave equation to describe quantum phenomena.
- 1927 Clinton Joseph Davisson (1881–1958) and Lester Halbert Germer (1896–1971) discover the undulatory behavior of electrons.
- 1920s soft magnetic alloys Mumetal and Permalloy invented.
- 1932 Louis Néel (1904–2000) suggests the existence of antiferromagnetism.
- 1930s Tokushichi Mishima (1893–1975) invents the permanent magnet alloy Alnico.
- 1932 Cornelius Jacobus Gorter (1907–1980) attempts to observe magnetic resonance and fails.
- 1935 Magnetophon: the first magnetic tape recorder.
- 1930s ferrites are studied.

- 1944 Evgeny K. Zavoisky (1907–1976) discovers magnetic resonance (electron paramagnetic resonance).
- 1946 two American groups discover nuclear magnetic resonance.
- 1956 magnetic hard disk introduced by IBM.
- 1956 first commercial video magnetic recorder.
- 1963 magnetic compact cassette introduced.
- 1973 Paul C. Lauterbur (1928) publishes an image obtained with nuclear magnetic resonance.
- 1981 the first precision map of the Earth magnetic field is made by the satellite MAGSAT.
- 1983 development of permanent magnet based on NdFeB.
- 1988 giant magnetoresistance (GMR) is reported by a group in Orsay led by A. Fert.
- 1997 first prototype of magnetic refrigerator is built.
- 1998 two-qubit quantum computer using nuclear magnetic resonance is demonstrated.
- 1998 quantum teleportation using nuclear magnetic resonance is demonstrated.

Glossary

A

Action at a distance: effect produced by a given agent at a distance, without physical contact.
Amber: fossil resin that becomes charged when frictioned with a piece of cloth or fur.
Animal magnetism: doctrine associated with the name of Franz Mesmer (1734–1815), who believed humans had a natural *magnetism* that could be used for therapeutic purposes.
Anisotropy, magnetic: the property that makes the *magnetization* of a sample prefer to point along a given direction.
Antiferromagnet: type of material where the *magnetic moments* of the atoms are antiparallel, leading to a near zero resulting moment.

C

Coercivity: magnitude of the *magnetic field* in the negative direction for which the *magnetization* reaches zero, after having reached its maximum value; in a *hysteresis curve*, the *coercivity* is given by half the width of the curve.
Compass: simple instrument formed of a magnetized needle that can turn, aligning itself with the Earth's *magnetic field*.

D

Declination, magnetic: angle between the direction of the *magnetic field* on the surface of the Earth and the direction of the meridian, or the true North-South direction.
Diamagnet: type of substance that is repelled by a *magnet*.
Dip: the same as *magnetic inclination*.
Domain: small region in a magnetized body (e.g. a *ferromagnet*), where the atomic *magnetic moments* are all pointing in the same direction. The *ferromagnet* normally contains many *domains* with moments along different directions, resulting in a total *magnetic moment* near zero.

From Lodestone to Supermagnets. Alberto P. Guimarães

ISBN: 3-527-40557-7

E

Electric charge: fundamental property of particles such as the *electron*; particles may have positive, negative or zero charge.

Electrolysis: chemical decomposition induced by a flow of electric current.

Electromagnet: device consisting of a coil wound on an iron armature, used to produce *magnetic fields* when electric current flows through the coil.

Electromagnetic wave: periodic magnetic and electric fields that propagate in space. Radio waves, light waves and X-rays are examples of *electromagnetic waves*.

Electron: particle with negative charge that is a constituent of the atom. It has a *magnetic moment* resulting from its *spin* and from its orbital motion.

Electrostatic machine: machine that charges objects, usually through friction.

Ether: fluid once thought to be necessary for the propagation of light, an idea abandoned since the beginning of the 20th century.

Exchange: phenomenon of quantum mechanical origin that induces an alignment of the electron magnetic moments, as e.g. in a *ferromagnet*.

F

Ferrite: iron compound of general formula MFe_2O_4, where M is a metal.

Field: property of the space that is described by a value of a variable associated with each point of space. One may speak of a field of temperatures, or velocities, when to each point a temperature or velocity may be assigned.

Field, magnetic: property of the space near a *magnet* or a wire where electric current flows, that may be observed as a turning force on a *compass* needle.

Ferromagnet: type of material where the *magnetic moments* of the atoms are aligned in parallel; a *ferromagnet* is strongly attracted by a magnet.

H

Hard magnetic material: material that has a large *coercivity*, and is useful for the manufacture of permanent magnets.

Hysteresis curve: closed curve of the *magnetization* versus *magnetic field*, used to characterize the properties of a magnetic material.

I

Inclination, magnetic: angle formed between the *magnetic field* on the surface of the Earth and the horizontal plane (the same as dip).
Induction: the effect of a magnet on a previously unmagnetized object, ordering it, and making this object act as another magnet.
Induction, electromagnetic: the phenomenon of the appearance of a voltage on a closed circuit that is crossed by a time-varying *magnetic field*.

L

Leyden jar: early type of capacitor made of a jar containing metal sheet or foil in contact with a central conductor, and covered by metal on the outside.
Line of field, or **line of force**: graphical representation of a *magnetic field*; the intensity of the field is represented by the number of lines per unit area.
Lodestone (or loadstone): earlier designation of naturally occurring magnetic mineral; magnetized *magnetite,* or *magnetite* that presents a *magnetic moment.*

M

Magnet: magnetized body; sample with a spontaneous *magnetic moment.*
Magnetar: a special type of neutron star, on which surface there is an ultra intense *magnetic field.*
Magnetism: phenomenon involving the effect of *magnetic fields,* the interaction between a *magnet* and other objects.
Magnetite: naturally occurring magnetic mineral, an iron oxide of formula Fe_3O_4.
Magnetization: a measure of the degree of magnetic order of a body, given by the total *magnetic moment* divided by the volume.
Magnetize: to turn into a magnet. A *magnetic field magnetizes* a *ferromagnet,* by producing an arrangement of *domains* preferentially aligned in the direction of the field, which results in a non-zero *magnetic moment.*

Magnetoresistance: phenomenon where the electrical resistance of a sample varies with an applied *magnetic field*.
Moment, magnetic: quantity that measures the intensity of the effects of a *magnet*; the *magnetic field* around a magnet is larger if the *magnetic moment* is bigger.

O

Order, magnetic: state of magnetic materials that present the *magnetic moments* disposed in some regular way, e.g. parallel, as in a *ferromagnet*.

P

Paramagnet: type of substance that contains magnetic moments that are not ordered; a paramagnet is attracted by a *magnet*.
Permeability: property of materials that essentially measures the degree of magnetization produced by an applied magnetic field.
Poles, magnetic: the two regions of the *magnet* where the *magnetic field* is more intense.

Q

Quantum mechanics: branch of physics that describes phenomena of the atomic and subatomic scale.

R

Rare earths: elements of the Periodic Table that often have large *magnetic anisotropy*, and are used as constituents of some *hard magnetic materials*. Usually include the lanthanides and the elements scandium and yttrium.

S

Soft magnetic material: material that presents a low *coercivity*, and is useful for the manufacture of transformer cores, magnetic shields and sensors.
Spin: intrinsic rotational momentum of some particles, as the *electron*.
Supermagnet: *hard magnetic material* or strong permanent *magnet* made from alloys or compounds usually containing the *rare earth* elements.

T

Tesla: unit of *magnetic field*. The field on the Earth surface is of the order of 0.00005 T.

Transformer: device used to convert AC voltage from one value to another; in its simplest form, it consists of an iron ring around which two coils are wound.

References

Chapter 1

1 Johann Wolfgang von Goethe, "Gott, Gemüt und Welt", *A Dictionary of Scientific Quotations*, Alan L. MacKay (Compiler), Peter Brian Medawar, Revised edition (July 1991), Adam Hilger, p. 105.
2 P. R. S. Moorey, *Ancient Mesopotamian Materials and Industries: the Archaeological Evidence*, Oxford University Press, Oxford, 1994, reprinted Eisenbrauns, Winona Lake, Indiana, 1999, p. 84.
3 R. Campbell Thompson, *A Dictionary of Assyrian Chemistry and Geology*, Clarendon Press, Oxford, 1936, p. 85.
4 *Akkadisches Handwörterbuch*, Heidelberger Akademie der Wissenschaften, Otto Harrassowitz, Wiesbaden, 1981, vol. III, p. 1123.
5 Erica Reiner, Ed., *The Assyrian Dictionary of the Oriental Institute of the University of Chicago*, Oriental Institute, Chicago, and J. J. Augustin Verlagsbuchhandlung, Glückstadt, 1989, vol. 17, p. 36.
6 J. Huehnergard, private communication (2000).
7 J. Huehnergard, private communication (2004).
8 Lucretius, "On the Nature of Things", *Great Books*, vol. 12, Encyclopaedia Britannica, Chicago, 1978, no. 906, p. 92.
9 A. C. Pearson, quoted by J. B. Kramer, "The Early History of Magnetism", Transactions of the Newcomen Society, vol. 14, 183–200 (1933–1934), p. 185.
10 J. B. Kramer, "The Early History of Magnetism", Transactions of the Newcomen Society, vol. 14, 183–200 (1933–1934), p. 186.
11 Pliny the Elder, *Natural History: a Selection*, Book XXVI, no. 127, Penguin Books, London, 1991, p. 359.
12 R. J. Forbes, "Extracting, Smelting and Alloying", in *A History of Technology*, eds. Charles Singer, E. J. Holmyard and A. R. Hall, vol. 1, Oxford University Press, Oxford, 1967, p. 594.
13 H. Hodges, *Technology in the Ancient World*, Penguin, Harmondsworth, 1971.
14 W. K. C. Guthrie, *A History of Greek Philosophy*, vol. I, Cambridge University Press, Cambridge, 1967, p. 33.
15 C. A. Ronan, *The Cambridge Illustrated History of the World's Science*, Cambridge University Press, Cambridge, 1984, p. 143.
16 J. Needham, quoted by P. Acot, *L'Histoire des Sciences*, Collection Que sais-je?, Presses Universitaires de France, Paris, 1999, p. 55.
17 W. K. C. Guthrie, *A History of Greek Philosophy*, vol. I, Cambridge University Press, Cambridge, 1967, p. 142.
18 P. Pellegrin, "Physique", p. 459, in Jacques Brunschwig, Geofffrey Lloyd and Pierre Pellegrin, *Le Savoir Grecque*, Flammarion, Paris, 1996, p. 463.
19 G. E. R. Lloyd, *Magic, Reason and Experience*, Cambridge University Press, Cambridge, 1993, p. 243.

From Lodestone to Supermagnets. Alberto P. Guimarães

ISBN: 3-527-40557-7

20 G. E. R. Lloyd, *Magic, Reason and Experience*, Cambridge University Press, Cambridge, 1993, p. 239.
21 Aristotle, Metaphysics, Book I, 981b, *Great Books*, vol. 8, Encyclopaedia Britannica, Chicago, 1978, p. 500.
22 Encyclopedia Britannica CD, Version 99 © 1994–1999.
23 P. H. Michel and C. Mugler, "Les Sciences dans le Monde Gréco-Romain", in R. Taton, *Histoire Générale des Sciences*, vol. I, Presses Universitaires de France, Paris, 1966, p. 208.
24 Herodotus, Hdt. II, 53, quoted by W. K. C. Guthrie, *A History of Greek Philosophy*, vol. I, Cambridge University Press, Cambridge, 1967, p. 371.
25 W. K. C. Guthrie, *A History of Greek Philosophy*, vol. I, Cambridge University Press, Cambridge, 1967, p. 371.
26 J. Frazer, *The Golden Bough*, London, 1890, quoted by R. S. Brumbaugh, *The Philosophers of Greece*, State University of New York Press, Albany, 1981.
27 W. K. C. Guthrie, *The Greek Philosophers – from Thales to Aristotle*, Methuen & Co, London, 1975, p. 24.
28 J. Barnes, *Early Greek Philosophy*, Penguin, London, 1987, p. 25.
29 W. K. C. Guthrie, *A History of Greek Philosophy*, vol. I, Cambridge University Press, Cambridge, 1967, p. 59.
30 C. B. Boyer and U. C. Merzbach, *A History of Mathematics*, 2nd edition, John Wiley, New York, 1989, p. 55.
31 W. K. C. Guthrie, *A History of Greek Philosophy*, vol. I, Cambridge University Press, Cambridge, 1967, p. 46; O. Neugebauer, *The Exact Sciences in Antiquity*, 2nd edition, Dover, New York, 1969, p. 142; D. R. Dicks, *Early Greek Astronomy*, Chapman and Hall, 1970, p. 43.
32 Simplicius, "Commentary on the Physics" 24.13–25, in Barnes, *Early Greek Philosophy*, Penguin, London, 1987, p. 74.
33 [Plutarch], "On the Scientific Beliefs of the Philosophers", 876 AB, quoted by J. Barnes, *Early Greek Philosophy*, Penguin, London, 1987, p. 79.
34 W. K. C. Guthrie, *A History of Greek Philosophy*, vol. I, Cambridge University Press, Cambridge, 1967, p. 151.
35 W. K. C. Guthrie, *A History of Greek Philosophy*, vol. I, Cambridge University Press, Cambridge, 1967, p. 181.
36 W. K. C. Guthrie, *A History of Greek Philosophy*, vol. I, Cambridge University Press, Cambridge, 1967, p. 213.
37 J. Barnes, *Early Greek Philosophy*, Penguin, London, 1987, p. 209.
38 J. Barnes, *Early Greek Philosophy*, Penguin, London, 1987, p. 210.
39 M. Kline, *Mathematical Thought from Ancient to Modern Times*, Oxford University Press, New York, 1972.
40 W. K. C. Guthrie, *A History of Greek Philosophy*, vol. I, Cambridge University Press, Cambridge, 1967, p. 224.
41 R. S. Brumbaugh, *The Philosophers of Greece*, State University of New York Press, Albany, 191, p. 151.
42 W. K. C. Guthrie, *The Greek Philosophers – from Thales to Aristotle*, Methuen & Co, London, 1975, p. 88.
43 D. Ross, *Aristotle*, Methuen & Co, London, 1974, p. 32.
44 Encyclopedia Britannica CD, Version 99 © 1994–1999.
45 C. C. Gillispie, *The Edge of Objectivity*, Princeton University Press, Princeton, 1973, p. 12.

46 R. S. Brumbaugh, *The Philosophers of Greece*, State University of New York Press, Albany, 1981, p. 194.
47 A. Koyré, *Galileo and Plato*, Journal of the History of Ideas, vol. 4, no. 4 (1943) 400, p. 421.
48 Aristotle, On the Soul, "The Works of Aristotle", vol. I, Encyclopaedia Britannica, Chicago 1952, p. 634.
49 J. Barnes, op. cit., p. 66.
50 W. K. C. Guthrie, *The Greek Philosophers- from Thales to Aristotle*, Methuen & Co, London, 1975, p. 66.
51 Pliny the Elder, Book XXXVII, no. 48, op. cit., p. 370.
52 Plato, on Ion, *Great Books*, vol. 7, Encyclopaedia Britannica, Chicago, 1952 p. 144.
53 W. K. C. Guthrie, *A History of Greek Philosophy*, vol. II, quoting *Quaestiones* A89, p. 232.
54 W. K. C. Guthrie, *A History of Greek Philosophy*, vol. II, quoting *Quaestiones*, A165, p. 426.
55 Galen, "On the natural faculties", *Great Books*, vol. 10, Encyclopaedia Britannica, Chicago 1952, p. 177
56 J. Needham, *Science and Civilisation in China*, vol. 4, part I, Cambridge University Press, Cambridge, 1972, p. 232.
57 C. A. Ronan, *The Cambridge Illustrated History of the World's Science*, Cambridge University Press, Cambridge, 1984, p. 161.
58 C. A. Ronan, *The Cambridge Illustrated History of the World's Science*, Cambridge University Press, Cambridge, 1984, p. 185.
59 C. A. Ronan, *The Cambridge Illustrated History of the World's Science*, Cambridge University Press, Cambridge, 1984, p. 141.
60 A. Haudricourt and J. Needham, "La Science Chinoise Antique", in R. Taton, *Histoire Générale des Sciences*, vol. I, Presses Universitaires de France, Paris, 1966, p. 198.
61 C. A. Ronan, *The Cambridge Illustrated History of the World's Science*, Cambridge University Press, Cambridge, 1984, p. 142.
62 J. Needham, *Science and Civilisation in China*, vol. 4, part I, Cambridge University Press, Cambridge, 1972, p. 31.
63 J. Needham, *Science and Civilisation in China*, vol. 4, part I, Cambridge University Press, Cambridge, 1972, p. 232.
64 J. Needham, *Science and Civilisation in China*, vol. 4, part I, Cambridge University Press, Cambridge, 1972, p. 232.
65 J. Needham, *Science and Civilisation in China*, vol. 4, part I, Cambridge University Press, Cambridge, 1972, p. 233.
66 J. Needham, *Science and Civilisation in China*, vol. 4, part I, Cambridge University Press, Cambridge, 1972, p. 262.
67 J. Needham, vol. 4, *Physics and Physical Technology*, Part II, Mechanical Engineering, 1974, p. 286.
68 J. Needham, *Science and Civilisation in China*, vol. 4, part I, Cambridge University Press, Cambridge, 1972, p. 230.
69 A. P. Guimarães, "Mexico and the Early History of Magnetism", Revista Mexicana de Física (2004).
70 J. B. Carlson, Science vol. 189 (1975) 753.
71 J. B. Carlson, Science vol. 189 (1975) 753.

72 Vincent H. Malmström, in *Encyclopaedia of the History of Science, Technology and Medicine in Non-Western Cultures*, Ed. Helaine Selin, Kluwer Academic, Dordrecht, 1997, p. 543.
73 V.H. Malmstrom, Nature vol. 259 (1976) 390.
74 Lucretius, "On the Nature of Things", *Great Books*, vol. 12, Encyclopaedia Britannica, Chicago, 1978, no. 1042, p. 94.
75 Pliny the Elder, in M.R. Cohen and I.E. Drabkin, *A Source Book in Greek Science*, Harvard University Press, Cambridge, 1966, p. 311.
76 M.R. Cohen and I.E. Drabkin, *A Source Book in Greek Science*, Harvard University Press, Cambridge, 1966, p. 312.
77 Pliny the Elder, *Natural History: a Selection*, Book II, no. 211, Penguin Books, London, 1991, p. 39.
78 R. Wallace, *"Amaze your Friends!" Lucretius on Magnets*, Greece & Rome, xliii, No. 2, October 1996, 178, p. 185.
79 Claudian, "Idyl V", quoted by Alfred Still, *Soul of Lodestone: the background of magnetical science*, Murray Hill Books, New York, 1946, p. 3.
80 R. Arnaldez, L. Massignon and A.P. Youschkevitch, "La Science Arabe", in R. Taton, *Histoire Générale des Sciences*, vol. I, Presses Universitaires de France, Paris, 1966, p. 445.

Chapter 2

1 G. Beaujouan, "La Science dans l'Occident Médiéval Chrétien", in R. Taton, *Histoire Générale des Sciences*, vol. I, Presses Universitaires de France, Paris, 1966, p. 583.
2 A.C. Crombie, *The History of Science from Augustine to Galileo*, vol. 2, Dover Publications, New York, 1995, p. 117.
3 A.C. Crombie, *The History of Science from Augustine to Galileo*, vol. 2, Dover Publications, New York, 1995, p. 126.
4 R. Arnaldez, L. Massignon and A.P. Youschkevitch, "La Science Arabe", in R. Taton, see Ref 1, p. 462.
5 C.A. Ronan, *The Cambridge Illustrated History of the World's Science*, Cambridge University Press, Cambridge, 1984, p. 229.
6 R. Arnaldez, L. Massignon and A.P. Youschkevitch, "La Science Arabe", in R. Taton, see Ref 1., p. 512.
7 R. Arnaldez, L. Massignon and A.P. Youschkevitch, "La Science Arabe", in R. Taton, see Ref 1., p. 498.
8 R.T. Merrill, M.W. McElhinny and P.L. McFadden, *The Magnetic Field of the Earth*, Academic Press, San Diego, 1998, p. 170.
9 J. Needham (a), op. cit., p. 252.
10 J.A. Smith, *Precursors to Peregrinus: The early history of magnetism and the mariners' compass in Europe*, Journal of Medieval History vol. 18 (1992) 21.
11 J.A. Smith, *Precursors to Peregrinus: The early history of magnetism and the mariners' compass in Europe*, Journal of Medieval History vol. 18 (1992) 21.
12 Alexander Neckam, "De Nominibus Utensilium", quoted by J.A. Smith, *Precursors to Peregrinus: The early history of magnetism and the mariner's compass in Europe*, J. of Medieval History vol. 18 (1992) 21, p. 34.
13 J.A. Smith, *Precursors to Peregrinus: The early history of magnetism and the mariner's compass in Europe*, Journal of Medieval History vol. 18 (1992) 21, p. 37.

13 J. A. Smith, *Precursors to Peregrinus: The early history of magnetism and the mariner's compass in Europe,* Journal of Medieval History vol. 18 (1992) 21, p. 37.
14 Debala Mitra, *Konarak,* published by the Director General, Archaeological Survey of India, New Delhi, 1976, p. 10, quoted by C. K. Majumdar, in "Magnetism: Past, Present and Future", in *Current Trends in Magnetism,* Eds. N. S. Satya Murthy and L. M. Rao, Indian Physics Association, Bombay, 1981, p. 1.
15 J. A. Smith, *Precursors to Peregrinus: The early history of magnetism and the mariner's compass in Europe,* Journal of Medieval History vol. 18 (1992) 21, p. 38.
16 J. A. Smith, *Precursors to Peregrinus: The early history of magnetism and the mariner's compass in Europe,* Journal of Medieval History vol. 18 (1992) 21.p. 24
17 J. A. Smith, *Precursors to Peregrinus: The early history of magnetism and the mariner's compass in Europe,* Journal of Medieval History vol. 18 (1992) 21. p. 46.
18 J. A. Smith, *Precursors to Peregrinus: The early history of magnetism and the mariner's compass in Europe,* Journal of Medieval History vol. 18 (1992) 21, p. 56.
19 Lynn Thorndike, *John of St. Amand on the Magnet,* ISIS, vol. 36 (1945), pp. 156–157, p. 156.
20 E. Grant (a), "Peter Peregrinus", in *Dictionary of Scientific Biographies,* Ed. C. C. Gillispie, Charles Scribner's Sons, New York, 1980.
21 Pierre de Marincourt (a), "The Letter of Peregrinus", in *Source Book in Medieval Science,* ed. E. Grant, Harvard University Press, Cambridge, 1974, p. 368.
22 J. A. Smith, *Precursors to Peregrinus: The early history of magnetism and the mariner's compass in Europe,* Journal of Medieval History vol. 18 (1992) 21, p. 73.
23 Pierre de Marincourt (a), p. 369.
24 Pierre de Marincourt (a), "The Letter of Peregrinus", in *Source Book in Medieval Science,* ed. E. Grant, Harvard University Press, Cambridge, 1974, p. 368.
25 St. Thomas Aquinas, "*Opera Omnia*", Antwep, 1612, vol. 8, *Quaestio Unica: de Anima, art. 1 (Utrum anima humana possit esse forma et hoc aliquid*), p. 437, quoted by W. J. King, *The Natural Philosophy of William Gilbert and His Predecessors,* Contributions from the Museum of History and Technology, Smithsonian Institution Bulletin 218, Washington, 1959, 122–139, p. 127.
26 J. D. Bernal, *The Extension of Man,* Paladin, Frogmore, 1972, p. 124.
27 Pierre de Marincourt (a); note on page 374.
28 Pierre de Marincourt (a), p. 375.
29 Pierre de Marincourt (a), p. 375.
30 Quoted by E. Grant (a), see Ref 20, p. 533.
31 E. Grant (a), see Ref 20 p. 532.
32 J. A. Smith, *Precursors to Peregrinus: The early history of magnetism and the mariner's compass in Europe,* Journal of Medieval History vol. 18 (1992) 21.
33 W. J. King, *The Natural Philosophy of William Gilbert and His Predecessors,* Contributions from the Museum of History and Technology, Smithsonian Institution Bulletin 218, Washington, 1959, 122–139, p. 129.
34 Robert Norman, *The New Attractive,* quoted by R. Harré, *The Method of Science,* Wykeham Publications, London, 1970, p. xi.
35 Robert Norman, *The New Attractive,* London, 1581, quoted by E. Zilsel, *The Origins of William Gilbert's Scientific Method,* Journal of the History of Ideas vol. 2, no. 1 (1941) 1, p. 23.
36 Robert Norman, *The New Attractive,* London, 1581, quoted by W. J. King, *The Natural Philosophy of William Gilbert and His Predecessors,* Contributions from the Museum of History and Technology, Smithsonian Institution Bulletin 218, Washington, 1959, 122–139, p. 124.
37 W. Gilbert (a), De Magnete, *Great Books,* vol. 28, Encyclopaedia Britannica, Chi-

37 W. Gilbert (a), De Magnete, *Great Books*, vol. 28, Encyclopaedia Britannica, Chicago 1978, p. 1.
38 W. Gilbert (a), op. cit., p. 3.
39 W. Gilbert (a), op. cit., p. 58.
40 W. Gilbert (a), op. cit., p. 24.
41 W. Gilbert (a), op. cit., p. 25.
42 W. Gilbert (a), op. cit., p. 24.
43 W. Gilbert (a), op. cit., p. 26.
44 Duane H. D. Roller, *The De Magnete of William Gilbert*, Menno Hertzberger, Amsterdam, 1959, p. 92.
45 W. Gilbert (a), op. cit., p. 31.
46 W. Gilbert (a), op. cit., p. 31.
47 W. Gilbert (a), op. cit., pp. 36–40.
48 W. Gilbert (a), op. cit., p. 53.
49 W. Gilbert (a), op. cit., p. 75.
50 W. Gilbert, quoted by Duane Roller
51 W. Gilbert (a), op. cit., p. 82.
52 W. Gilbert (a), op. cit., p. 103.
53 W. Gilbert (a), op. cit., p. 116.
54 W. Gilbert (a), op. cit., p. 117.
55 W. Gilbert (a), op. cit., p. 110.
56 W. Gilbert (a), op. cit., p. 112.
57 W. Gilbert (a), op. cit., p. 117.
58 W. Gilbert (a), op. cit., p. 117.
59 W. Gilbert (a), op. cit., p. 77.
60 W. Gilbert (a), op. cit., p. 79.
61 A. Kircher, *Magnes, sives De Arte Magnetica*, Rome, 1641, p. 479, quoted by M. R. Baldwin, *Magnetism and the Anti-Copernican Polemic*, Journal of the History of Astronomy vol. 16 (1985) 155, p. 159.
62 G. Freudenthal (a), *Theory of Matter and Cosmology in William Gilbert's De magnete*, ISIS, vol. 74 (1983) 22–37, p. 25.
63 G. Freudenthal (a), op. cit. p. 28.
64 G. Freudenthal (a), op. cit. p. 34.
65 G. Freudenthal (a), op. cit. p. 33.
66 W. Gilbert (a), op. cit., p. 115.
67 J. A. Bennett (a), *Cosmology and the Magnetical Philosophy*, Journal of the History of Astronomy vol. 12 (1981) 165, p. 165.
68 Johannes Kepler, *Gesammelte Werke*, ed. by Walther von Dyck and Max Caspar, Munich, 1937–75, xvi, 86, Letter to Johann Georg Brengger of 30 November 1607, quoted by M. R. Baldwin, *Magnetism and the Anti-Copernican Polemic*, Journal of the History of Astronomy vol. 16 (1985) 155, p. 156.
69 W. Gilbert (a), op. cit., p. 104.
70 W. Gilbert (a), op. cit., p. 49.
71 E. Zilsel (a), *The Origins of William Gilbert's Scientific Method*, Journal of the History of Ideas vol. 2, no. 1 (1941) 1, p. 31.
72 E. Zilsel (a), op. cit., p. 15.
73 Duane H. D. Roller, *The De Magnete of William Gilbert*, Menno Hertzberger, Amsterdam, 1959, p. 91.
74 Duane H. D. Roller, *The De Magnete of William Gilbert*, Menno Hertzberger, Amsterdam, 1959, p. 50.

75 John Dryden, "Poems", ("To My Honour'd Friend D[r] Charleton"), quoted by Patricia Fara, *Sympathetic Attractions*, Princeton University Press, Princeton, 1996.
76 Augustine, *De Genesi ad Litteram*, chapter 21, quoted by A.C. Crombie (a), op. cit., vol. 1, p. 75.
77 C.A. Ronan, *The Cambridge Illustrated History of the World's Science*, Cambridge University Press, Cambridge, 1984, p. 260.
78 J.A. Bennett (a), *Cosmology and the Magnetical Philosophy, 1640–1680*, Journal of the History of Astronomy vol. 12 (1981) 165, p. 166.
79 J.A. Bennett (a), op. cit., p. 171.
80 A.C. Crombie, *The History of Science from Augustine to Galileo*, vol. 2, Dover Publications, New York, 1995, p. 145.
81 A.C. Crombie, *The History of Science from Augustine to Galileo*, vol. 2, Dover Publications, New York, 1995, p. 138.
82 Galileo Galilei, *Dialogues Concerning the Two New Sciences*, Fourth Day, Great Books 28, Encyclopedia Britannica, Chicago, 1978, p.238.
83 W.R. Shea, *Galileo"s Intellectual Revolution*, 2nd ed., Science History Publications, New York, 1977, p. 167.
84 Galileo Galilei (a), Third Day, p. 400.
85 Galileo Galilei (a), Third Day, p. 400.
86 Galileo Galilei (a), *Dialogues Concerning the Two Chief World Systems – Ptolemaic & Copernican*, "Third Day", translated by Stillman Drake, 2nd ed., University of California Press, Berkeley, 1967, p. 413.
87 W. Barlowe, *Magneticall Advertisement*, London, 1616, quoted by E. Zilsel, *The Origins of William Gilbert's Scientific Method*, Journal of the History of Ideas vol. 2, no. 1 (1941) 1, p. 29.
88 Galileo Galilei, quoted by J. Bronowski, *The Ascent of Man*, Little, Brown and Company, Boston, 1973, p. 204.
89 A. Koyré, *Galileo and Plato*, Journal of the History of Ideas, vol. 4, no. 4 (1943) 400, p. 404.
90 Galileo Galilei, quoted by J. Bronowski, *The Ascent of Man*, Little, Brown and Company, Boston, 1973, p. 217.
91 R. Descartes, 1638, quoted by A.C. Crombie (a), op. cit., vol. 2, p. 171.
92 Galileo Galilei (a), *Dialogues Concerning the Two Chief World Systems – Ptolemaic & Copernican*, "Third Day", translated by Stillman Drake, 2nd ed., University of California Press, Berkeley, 1967, p. 406.
93 M. Kline, *Mathematics and the Search of Knowledge*, Oxford University Press, New York, 1985, p. 95.
94 H. Dingle, "Physics in the Eighteenth Century", in *Natural Philosophy Through the Eighteenth Century*, Taylor and Francis, London, 1972, p. 28.

Chapter 3

1 Max Planck, "The Unity of the Physical World-Picture", in *Physical Reality, Essays on Twentieth-Century Physics*, Ed. Stephen Toulmin, Harper & Row, New York, 1970, pp. 1–27, p. 3.
2 Pliny the Elder, *Natural History: a Selection*, Book XXXVII, no. 48, Penguin Books, London, 1991, p. 370.

3 Duane H. D. Roller, *The De Magnete of William Gilbert*, Menno Hertzberger, Amsterdam, 1959, p. 22.
4 Plato, in "Meno", [80a], *Great Books*, vol. 7, Encyclopaedia Britannica, Chicago 1978, p. 179.
5 Aristotle, *History of Animals*, 620b [18], *Great Books*, vol. 9, Encyclopaedia Britannica, Chicago 1978, p. 146.
6 George Sarton, *Introduction to the History of Science*, vol. I, Robert E. Krieger Publishing Company, Malabar, 1927, Reprinted 1975, p. 241.
7 Chau H. Wu, *Electric Fish and the Discovery of Animal Electricity*, American Scientist, vol. 72, November-December 1984, pp. 598–607, p. 601.
8 Niccolo Cabeo, "Philosophia Magnetica" (1629), vol. II, p. 21, quoted by A. Wolf, *History of Science, Technology and Philosophy in the 17th Century*, p. 303.
9 A. Wolf, *History of Science, Technology and Philosophy in the 16th and 17th Century*, vol. 1, 2nd edition, George Allen & Unwin, London, 1962, p. 303.
10 A. Wolf, *History of Science, Technology and Philosophy in the 16th and 17th Century*, vol. 1, 2nd edition, George Allen & Unwin, London, 1962, p.303.
11 E. Bauer, "L'Electricité et le Magnetisme au XVIIIe siècle", in R. Taton (a), *Histoire Générale des Sciences*, vol. II, Presses Universitaires de France, Paris, 1958, p. 523.
12 E. Bauer, "L'Electricité et le Magnetisme au XVIIIe siècle", in R. Taton (a), *Histoire Générale des Sciences*, vol. II, Presses Universitaires de France, Paris, 1958, p. 523.
13 E. Bauer, "L'Electricité et le Magnetisme au XVIIIe siècle", in R. Taton (a), *Histoire Générale des Sciences*, vol. II, Presses Universitaires de France, Paris, 1958, p. 525.
14 Bernard Cohen, in "Benjamin Franklin", Lives in Sciences, a Scientific American Book, Simon and Schuster, New York, 1957 (reprinted from Scientific American, August 1948), p. 117.
15 A. Wolf, *History of Science, Technology and Philosophy in the 18th Century*, vol. 1, 2nd edition, George Allen & Unwin, London, 1962, p. 230; Paul Fleury Mottelay, Bibliographical History of Electricity and Magnetism, Arno Press, New York, 1975, p. 152.
16 A. Wolf, *History of Science, Technology and Philosophy in the 18th Century*, vol. 1, 2nd edition, George Allen & Unwin, London, 1962, p. 234.
17 E. Bauer, "L'Electricité et le Magnetisme au XVIIIe siècle", in R. Taton (a), *Histoire Générale des Sciences*, vol. II, Presses Universitaires de France, Paris, 1958, p. 521.
18 A. Wolf, *History of Science, Technology and Philosophy in the 18th Century*, vol. 1, 2nd edition, George Allen & Unwin, London, 1962, p. 224.
19 A. Wolf, *History of Science, Technology and Philosophy in the 18th Century*, vol. 1, 2nd edition, George Allen & Unwin, London, 1962, p. 222.
20 Laura Bossi, "L'Âme Electrique", *L'Âme au Corps, Arts et Sciences 1793–1993*, Ed. Jean Clair, Galimard, Paris, 1994, pp. 160–180, p. 163.
21 T. M. Brown, "Galvani", in *Dictionary of Scientific Biographies*, Ed. C. C. Gillispie, Charles Scribner's Sons, New York, 1980, p. 268.
22 J. L. Heilbron, "Volta", in *Dictionary of Scientific Biographies*, Ed. C. C. Gillispie, Charles Scribner's Sons, New York, 1980, p. 76.
23 M. Godwin Shelley, *Frankenstein or the Modern Prometheus*, edited by M. K. Jozepth, Oxford University Press, Oxford, 1969, quoted by Laura Bossi, "L'Âme

Electrique“, *L'Âme au Corps, Arts et Sciences 1793–1993*, Ed. Jean Clair, Galimard, Paris, 1994, pp. 160–180, p. 169.
24 J.L. Heilbron, “Volta“, in *Dictionary of Scientific Biographies*, Ed. C.C. Gillispie, Charles Scribner's Sons, New York, 1980, p. 69.
25 J.L. Heilbron, “Volta“, in *Dictionary of Scientific Biographies*, Ed. C.C. Gillispie, Charles Scribner's Sons, New York, 1980, p. 76.
26 A. Wolf, *History of Science, Technology and Philosophy in the 18th Century*, vol. 1, 2nd edition, George Allen & Unwin, London, 1962, p. 256.
27 J.L. Heilbron, “Volta“, in *Dictionary of Scientific Biographies*, Ed. C.C. Gillispie, Charles Scribner's Sons, New York, 1980, p. 79.
28 S. Sambursky, *Physical Thought: from the Presocratics to the Quantum Physicists*, Pica Press, New York, 1974.
29 W.K.C. Guthrie, *The Greek Philosophers – from Thales to Aristotle*, Methuen & Co, London, 1967, p. 36.
30 Roger Boscovich, *Theoria Philosophiae Naturalis*, quoted in *Theories of Everything*, J.D. Barrow, Clarendon Press, Oxford, 1991, p. 17.
31 Steve Adams, *A theory of everything*, New Scientist vol. 161 (2174) (1999) A1.
32 C.F. von Weizsäcker, “The World-View of Physics“, 1952, translated by Marjorie Grene, from the 4th German edition, 1949, pg. 30, quoted by W.K.C. Guthrie, *A History of Greek Philosophy*, vol. I, Cambridge University Press, Cambridge, 1967, p. 70.
33 Immanuel Kant, *Metaphysical Foundations of Natural Science*, Bobbs-Merrill, Indianapolis, 1970, p. 93, quoted by Timothy Shanahan, “Kant, *Naturphilosophie*, and Oersted's Discovery of Electromagnetism: a Reassessment“, Studies in History and Philosophy of Science vol. 20 (1989) 287–305, p. 294.
34 Philosophical Transactions, London, vol. XI, pp. 647–653, (1676) p. 647.
35 Philosophical Transactions, London, vol. XXXIX, (1735) pp. 74 and 75.
36 Hans Christian Andersen, quoted by B. Dibner, *Oersted and the Discovery of Electromagnetism*, Blaisdell Publishing Company, New York, 1962, p. 64.
37 Timothy Shanahan, “Kant, *Naturphilosophie*, and Oersted's Discovery of Electromagnetism: a Reassessment“, Studies in History and Philosophy of Science vol. 20 (1989) 287–305.
38 Timothy Shanahan, “Kant, *Naturphilosophie*, and Oersted's Discovery of Electromagnetism: a Reassessment“, Studies in History and Philosophy of Science vol. 20 (1989) 287–305, p. 298.
39 Pierre Thuiller, “De la Philosophie à l'Électromagnetisme: le Cas Oersted“, *Recherche*, vol. 21 (1990) 344–350.
40 Arthur Eddington, quoted by M. Kline, *Mathematics and the Search of Knowledge*, Oxford University Press, New York, 1985, p. 217.
41 L. Pearce Williams, “Oersted“, in *Dictionary of Scientific Biographies*, Ed. C.C. Gillispie, Charles Scribner's Sons, New York, 1980, p. 185.
42 Robert C. Stauffer, “Speculation and Experiment in the Background of Oersted's Discovery of Electromagnetism“, ISIS vol. 48 (1953) 33–50, p. 41.
43 Timothy Shanahan, “Kant, *Naturphilosophie*, and Oersted's Discovery of Electromagnetism: a Reassessment“, Studies in History and Philosophy of Science vol. 20 (1989) 287–305, p. 287.
44 H.C. Oersted, “Thermo-electricity“, in Edinburgh Encyclopaedia (1830), XVIII, 573–589, quoted by L. Pearce William, “Oersted”, in *Dictionary of Scientific Biographies*, Ed. C.C. Gillispie, Charles Scribner's Sons, New York, 1980.

45 H.C. Oersted, "Annals of Philosophy", 1820, in S. Sambursky, *Physical Thought: from the Presocratics to the Quantum Physicists*, Pica Press (New York 1974) p. 379.
46 H.C. Oersted, "Annals of Philosophy", 1820, quoted by S. Sambursky, Physical *Thought: from the Presocratics to the Quantum Physicists*, Pica Press, New York, 1974, p. 380.
47 L. Pearce William, "Oersted", in *Dictionary of Scientific Biographies*, Ed. C.C. Gillispie, Charles Scribner's Sons, New York, 1980, p. 185.
48 L. Pearce Williams, "What Were Ampère's Earliest Discoveries in Electrodynamics?", ISIS vol. 74 (1983) 492–508, p. 507.
49 L. Pearce Williams, "What Were Ampère's Earliest Discoveries in Electrodynamics?", ISIS vol. 74 (1983) 492–508.
50 Hans Christian Oersted, *Scientific Papers*, ed. K. Mayer, Copenhagen, 1920, p. cxiv, quoted by L. Pearce Williams, "What Were Ampère's Earliest Discoveries in Electrodynamics?", ISIS vol. 74 (1983) 492–508, p. 494.
51 L. Pearce Williams, "What Were Ampère's Earliest Discoveries in Electrodynamics?", ISIS vol. 74 (1983) 492–508, p. 507.
52 A.M. Ampère, "Recueil d'observations électrodynamiques" (1822), in S. Sambursky, *Physical Thought: from the Presocratics to the Quantum Physicists*, Pica Press, New York, 1974, p. 385.
53 J.R. Hofmann, "Ampère, Electrodynamics and Experimental Evidence", Osiris 2nd series, vol. 3 (1987) 45–76, p. 48.
54 M. Faraday, Great Books vol. 45, Encyclopaedia Britannica, Chicago 1978, p. 265, Nov. 24, 1831.
55 M. Faraday, Great Books vol. 45, Encyclopaedia Britannica, Chicago 1978, p. 265, Nov. 24, 1831.
56 L. Pearce Williams, "Faraday", in *Dictionary of Scientific Biographies*, Ed. C.C. Gillispie, Charles Scribner's Sons, New York, 1980, p. 528.
57 D.K.C. MacDonald, *Faraday, Maxwell, and Kelvin*, Anchor Books, New York, 1964, quoted by G.L. Verschuur, *Hidden Attraction: the History and Mystery of Magnetism*, Oxford University Press, New York 1993, p. 82.
58 J.R. Hofmann, "Ampère, Electrodynamics and Experimental Evidence", Osiris 2nd series, vol. 3 (1987) 45–76, p. 76.
59 L. Pearce Williams, "Faraday", in *Dictionary of Scientific Biographies*, Ed. C.C. Gillispie, Charles Scribner's Sons, New York, 1980, p. 530.
60 R.J. Forbes and E.J. Dijksterhuis, *A History of Science and Technology*, vol. 2, Penguin Books, Harmondsworth, 1963, p. 456.
61 Miroslav Holub, "Magnetism", Poetry Review vol. 85 (1995) 29.
62 Mitchell Wilson, "Joseph Henry", in *Lives of Science*, pg. 141, a Scientific American Book, Simon and Schuster, New York 1957 (reprinted from Scientific American, July 1954) pg. 148.
63 J.R. Hofmann, "Ampère, Electrodynamics and Experimental Evidence", Osiris 2nd series, vol. 3 (1987) 45–76, p. 68.
64 H.S. Lipson, *The Great Experiments in Physics*, Oliver & Boyd, Edinburgh, 1968, p. 103.
65 E.J. Hobsbawn, *Industry and Empire, The Pelican Economic History of Britain, Volume 3, From 1750 to the Present Day*, Penguin Books, Harmondsworth, 1969, p. 173.

66 Duane H.D. Roller, *The De Magnete of William Gilbert*, Menno Hertzberger, Amsterdam, 1959, p. 27.
67 W. Gilbert, De Magnete, *Great Books*, vol. 28, Encyclopaedia Britannica, Chicago 1978, p. 20.
68 Albert L. Caillet, *Manuel Bibliographique des Sciences Psychiques ou Occultes*, Tome II, Reprinted by B. de Graaf, Nieuwkoop, 1964, p. 198.
69 E.G. Boring, "Hypnotism", in *A History of Experimental Psychology*, ed. R.M. Elliott, 2nd ed., Prentice-Hall, New Jersey, 1957, pp. 116–133, p. 116.
70 A. Wolf, *History of Science, Technology and Philosophy in the 16th and 17th Century*, vol. 1, 2nd edition, George Allen & Unwin, London, 1962, p.298.
71 Andrew Steptoe, *Mozart, Mesmer and 'Così Fan Tutte'*, Music & Letters, vol. 67, 1986, pp. 248–255, p. 252.
72 Heinz Schott, "Neurogamies. De la Relation Entre Mesmerism, Hypnose et Psychanalyse", *L'Âme au Corps, Arts et Sciences 1793–1993*, Ed. Jean Clair, Galimard, Paris, 1994, pp. 142–153, p. 153.

Chapter 4

1 Lucretius, *The Great Books* vol. 12, Encyclopaedia Britannica, Chicago,1978, p. 93.
2 Plato, "Timaeus", *The Great Books* vol. 7, Encyclopaedia Britannica, Chicago, 1978, p. 471.
3 Aristotle, Anal. Post 243a 16, quoted in S. Sambursky, *The Physical World of Late Antiquity*, Princeton University Press, Princeton 1962, p. 99.
4 Aristotle, "On Dreams", *The Great Books* vol. 8, Encyclopaedia Britannica, Chicago, 1978, p. 703.
5 Theon of Smirna, Expos. Rer. Math. 50, 22, in S. Sambursky, *The Physical World of Late Antiquity*, Princeton University Press, Princeton 1962, p. 101.
6 S. Sambursky, *The Physical World of Late Antiquity*, Princeton University Press, Princeton 1962, p. 102.
7 R. Descartes, "Principes de Philosophie", quoted by R. Lenoble, "Le Magnetisme et l'Électricité", in R. Taton, *Histoire Générale des Sciences*, vol. II, Presses Universitaires de France, Paris, 1958, pp. 333–4.
8 Isaac Newton, quoted in Arthur Koestler, *The Sleepwalkers*, Penguin Books, Harmondsworth, 1973, p. 344.
9 A. Haudricourt and J. Needham, "La Science Chinoise Antique", in R. Taton, *Histoire Générale des Sciences*, ed. R. Taton, vol. I, Part I, Chapter V, Presses Universitaires de France, Paris 1966, p. 198.
10 Francis Bacon, "Novum Organum" BII, *The Great Books* vol. 30, Encyclopaedia Britannica, Chicago, 1978, p. 165.
11 Johannes Kepler, "Epitome of Copernican Astronomy", IV, *The Great Books* vol. 16, Encyclopaedia Britannica, Chicago, 1978, p. 898.
12 Johannes Kepler, "Epitome of Copernican Astronomy", IV, *The Great Books* vol. 16, Encyclopaedia Britannica, Chicago, 1978, p. 934.
13 W.K.C. Guthrie, *A History of Greek Philosophy*, vol. I, Cambridge University Press, Cambridge, 1967, p. 271.
14 W.K.C. Guthrie, *A History of Greek Philosophy*, vol. I, Cambridge University Press, Cambridge, 1967, p. 136.

15 W.K.C. Guthrie, *A History of Greek Philosophy*, vol. I, Cambridge University Press, Cambridge, 1967, p. 466.
16 S. Sambursky, *Physical Thought: from the Presocratics to the Quantum Physicists*, Pica Press, New York, 1974, p. 248.
17 J.D. North, *The Measure of the Universe: A History of Modern Cosmology*, Dover, New York, 1990, p. 33.
18 Christiaan Huygens, "Treatise on Light", Ch. I, *The Great Books* vol. 34, Encyclopaedia Britannica, Chicago, 1978, p. 559.
19 R. Descartes, "Meteorology" (1637), in *The Science of Matter*, Ed. M.P. Crosland, Penguin, Harmondsworth 1971, p. 71.
20 Johannes Kepler, "Epitome of Copernican Astronomy", IV, *The Great Books* vol. 16, Encyclopaedia Britannica, Chicago, 1978, p. 857.
21 Isaac Newton, "Optics", *The Great Books* vol. 34, Encyclopaedia Britannica, Chicago 1978, pgs. 520, 521.
22 J.D. North, *The Measure of the Universe: A History of Modern Cosmology*, Dover, New York, 1990, p. 29.
23 R.S. Shankland, *The Michelson-Morley Experiment*, Scientific American, November 1964, p. 107.
24 Johannes Kepler, "Introduction to Astronomia Nova" (1609), quoted in Arthur Koestler, *The Sleepwalkers*, Penguin Books, Harmondsworth, 1973 p. 342.
25 Francis Bacon, "Novum Organum" BII, *The Great Books* vol. 30, Encyclopaedia Britannica, Chicago 1978, p. 176.
26 Quotation from Newton's recollections, Westfall, op. cit., p. 39.
27 R.S. Westfall, *The Life of Isaac Newton*, Cambridge University Press, Cambridge, 1993, pp. 51, 308.
28 I. Bernard Cohen, *The Birth of a New Physics*, Penguin, London, 1992, p. 238.
29 J. Henry, "Newton, Matter and Magic", in *Let Newton Be*, Ed. J. Fauvel, R. Flood, M. Shortland, R. Wilson, Oxford Press, Oxford, 1989, p. 144.
30 Isaac Newton, *Opticks, or a treatise of the reflections, refractions, inflections & colours of light*, New York, 1952, based on the 4th edition, London, 1730, pp. 375–6, quoted by R.W. Home, *Electricity and Experimental Physics in Eighteenth-Century Europe*, Variorum, Hampshire, 1992, p. 256.
31 Isaac Newton, "Optics", *The Great Books* vol. 34, Encyclopaedia Britannica, Chicago, 1978, p. 522.
32 R.W. Home, *Electricity and Experimental Physics in Eighteenth-Century Europe*, Variorum, Hampshire, 1992, p. 258.
33 R.W. Home, *Electricity and Experimental Physics in Eighteenth-Century Europe*, Variorum, Hampshire, 1992, pp. 260–261.
34 Isaac Newton, "Optics", *The Great Books* vol. 34, Encyclopaedia Britannica, Chicago, 1978, p. 531.
35 M.B. Hesse, *Forces and Fields: the Concept of Action at a Distance in the History of Physics*, Greenwood Press, Westport, 1962, p. 157.
36 M.B. Hesse, op.cit., p. 166.
37 Michael Faraday, *The Great Books* vol. 45, Encyclopaedia Britannica, Chicago, 1978, Twenty-eighth series, 1851, p. 759.
38 Herbert Kondo, "Michael Faraday", in *Lives of Science*, p. 127, a Scientific American Book, Simon and Schuster, New York 1957 (reprinted from Scientific American, October 1953) p. 138.
39 M. Faraday, "Experimental Researches", quoted by Herbert Kondo, p. 138.

40 James Clerk Maxwell, Proc. Roy. Inst. vol. VII, *The Scientific Papers of James Clerk Maxwell*, Ed. W.D. Niven, Dover, New York, 1965, part II, p. 311.
41 Quoted by Herbert Kondo, "Michael Faraday", in *Lives of Science*, p. 127, a Scientific American Book, Simon and Schuster, New York 1957 (reprinted from Scientific American, October 1953), p. 139.
42 J.R. Newman, "James Clerk Maxwell", in *Lives of Science*, p. 155, a Scientific American Book, Simon and Schuster, New York, 1957 (reprinted from Scientific American, June 1955), p. 172.
43 H. Hertz, "Gesammelte Werke", vol. 1, 1895, quoted in S. Sambursky, *Physical Thought: from the Presocratics to the Quantum Physicists*, Pica Press, New York, 1974, p. 461.
44 James Clerk Maxwell, *The Scientific Papers of James Clerk Maxwell*, Ed. W.D. Niven, Dover, New York, 1965, part II, p. 320.
45 A. Einstein, "Autobiographical Notes", in *Albert Einstein: Philosopher-Scientist*, I, Ed. P.A. Schilpp, Cambridge University Press, London, 1969, p. 33.
46 A. Einstein and L. Infeld, *The Evolution of Physics*, Cambridge University Press, 1938, quoted in *The Science of Matter*, Ed. M.P. Crosland, Penguin, Harmondsworth, 1971, p. 343.
47 Laurie M. Brown, ed., *Renormalization from Lorentz to Landau (and Beyond)*, Springer-Verlag, New York, 1993, p. 43.
48 M. Antoniazzi and G. Giuliani, "Campi, onde e particelle nei manuali di elletromagnetismo: un esempio di stratificazione concettuale e ontologica", Giornale di Fisica vol. 38 (1997) 87.
49 S. Weinberg, quoted by S.Y. Auyang, *How is Quantum Field Possible?*, Oxford University Press, New York, 1995.S

Chapter 5

1 R.M. Bozorth, *The Physical Basis of Ferromagnetism*, Bell Systems Technical Journal vol. 19 (1940) 1, p. 2, quoted by Stephen T. Keith and Pierre Quédec, "Magnetism and Magnetic Materials", in *Out of the Crystal Maze, Chapters from the History of Solid-State Physics*, Eds. Lillian Hoddeson, Ernest Braun, Jürgen Teichmannn, Spencer Weart, Oxford University Press, New York, 1992, p. 422.
2 G. Sarton, *A History of Science*, G. Cumberlege, London, 1953, p. 253.
3 Simplicius, Commentary on the Physics, quoted by J. Barnes, *Early Greek Philosophy*, Penguin, London, 1987, p. 248.
4 Simplicius, Commentary on the Heavens, quoted by J. Barnes, *Early Greek Philosophy*, Penguin, London, 1987, p. 247.
5 *The Science of Matter*, Ed. M.P. Crosland, Penguin, Harmondsworth, 1971, p. 40.
6 *The Science of Matter*, Ed. M.P. Crosland, Penguin, Harmondsworth, 1971, p. 41.
7 René Descartes, "Discourse on Method, Optics, Geometry and Meteorology", translated by P.J. Olscamp, Bobbs-Merrill, 1965, pp. 264–5, in *The Science of Matter*, Ed. M.P. Crosland, Penguin, Harmondsworth, 1971, p. 71.
8 René Descartes, "Philosophical Writings", translated by E. Anscombe and P.T. Geach, Nelson, 1954, p. 207, in *The Science of Matter*, Ed. M.P. Crosland, Penguin, Harmondsworth, 1971, p. 71.

9 Robert Boyle, "The Sceptical Chymist", 1661, modern edn., Dent, 1911, pp. 30–31, in *The Science of Matter*, Ed. M.P. Crosland, Penguin, Harmondsworth, 1971, p. 74.
10 J. Dalton, in *The Science of Matter*, Ed. M.P. Crosland, Penguin, Harmondsworth 1971, p. 202.
11 E.R. Scerri, *The Evolution of the Periodic System*, Scientific American, September 1998, p. 56.
12 J.J. Thomson, in *The Science of Matter*, Ed. M.P. Crosland, Penguin, Harmondsworth 1971, p. 356.
13 R. Rhodes, *The Making of the Atomic Bomb*, Penguin Books, London, 1986, p. 49.
14 E. Rutherford, quoted by Victor Guillemin, *The Story of Quantum Mechanics*, Charles Scribner's Sons, New York, 1968, p. 33.
15 Ernest Rutherford, quoted by Freeman Dyson, *Infinite in All Directions*, Penguin Books, London, 1988, p. 41.
16 Helge Kragh, *Quantum Generations, a History of Physics in the Twentieth Century*, Princeton University Press, Princeton, 1999, p. 3.
17 Helmut Rechenberg, "Quanta and Quantum Mechanics", in *Twentieth Century Physics*, vol. I, eds. Laurie M. Brown, Abraham Pais and Brian Pippard, Institute of Physics Publishing, Bristol, 1995, p. 150.
18 J.L. Heilbron, *The Dilemmas of an Upright Man – Max Planck as Spokesman for German Science*, University of California Press, Berkeley, 1986, p. 34.
19 J.L. Heilbron, *The Dilemmas of an Upright Man – Max Planck as Spokesman for German Science*, University of California Press, Berkeley, 1986, p. 98.
20 J.L. Heilbron, *The Dilemmas of an Upright Man – Max Planck as Spokesman for German Science*, University of California Press, Berkeley, 1986, p. 24.
21 A. Einstein, 'On the Investigation of the State of the Ether in the Magnetic Field', 1895, mentioned by Lewis Pyenson, *The Young Einstein: the Advent of Relativity*, Adam Hilger Ltd., Bristol, 1985, p. 8.
22 Abraham Pais, *Subtle is the Lord, the Science and Life of Albert Einstein*, Oxford University Press, Oxford, 1982, p. 245.
23 Time Magazine, New York, December 31, 1999.
24 A. Eucken, Anhang *Die Theorie der Strahlung und Quanten*, Knapp, Halle, 1914, p. 373, through Helmut Rechenberg, "Quanta and Quantum Mechanics", in *Twentieth Century Physics*, vol. I, eds. Laurie M. Brown, Abraham Pais and Brian Pippard, Institute of Physics Publishing, Bristol, 1995, p. 146.
25 R.H. Stuewer, *History and Physics*, Science & Education, vol. 7 (1998) pp. 13–30, p. 20.
26 Abraham Pais, "Introducing Atoms and Their Nuclei", in *Twentieth Century Physics*, vol. I, eds. Laurie M. Brown, Abraham Pais and Brian Pippard, Institute of Physics Publishing, Bristol, 1995, p. 95.
27 J.J. Thomson, *Structure of Light*, Cambridge, 1925, p. 15, quoted by B.R. Wheaton, *The Tiger and the Shark, Empirical Roots of Wave-Particle Dualism*, Cambridge University Press, Cambridge, 1992, p. 306.
28 Felix Bloch, Physics Today, vol. 29, December 1976, p. 23.
29 P.A.M. Dirac, *The Evolution of the Physicist's Picture of Nature*, Scientific American, May 1963, pp. 45–53.
30 P.A.M. Dirac, *The Development of Quantum Theory – J. Robert Oppenheimer Memorial Prize Acceptance Speech*, Gordon and Breach, New York, 1971, p. 66.

31 Werner Heisenberg, *Physics and Philosophy*, London, 1962, p. 77.
32 M. Tegmark and J. A. Wheeler, *100 Years of Quantum Mysteries*, Scientific American, February 2001, p. 54.
33 *The Born-Einstein Letters*, Irene Born, trans., Walker, New York, p. 158, quoted by N. David Mermin, "Spooky Actions at a Distance", The Great Ideas Today, Encyclopaedia Britannica, Chicago, 1988, pp. 2–53, p. 11.
34 E. Schrödinger, 'Discussion of the Probability Relations Between Separated Systems', Proceedings of the Cambridge Philosophical Society, vol. 31 (1935) pp. 555–563, quoted by Jeffrey Bub, 'Indeterminacy and Entanglement: The Challenge of Quantum Mechanics', British Journal of Philosophy of Science 51 (2000) pp. 597–615, p. 604.
35 A. Zeilinger, *Quantum Teleportation*, Scientific American, April 2000, p. 32.
36 M. Tegmark and J. A. Wheeler, *100 Years of Quantum Mysteries*, Scientific American, February 2001, p. 54.
37 R. Feynman, "The Character of Physical Law", 1965, p. 129, quoted by J. T. Cushing, *Philosophical Concepts in Physics*, Cambridge University Press, Cambridge, 2000, p. 317.
38 Albert Einstein, *Out of My Later Years*, quoted by Victor Guillemin, *The Story of Quantum Mechanics*, Charles Scribner's Sons, New York, 1968, p. 263.
39 John Stuart Mill, *A System of Logic*, Longmans, Green and Co., London, 8th edition, 1936, p. 213.
40 Pierre-Simon Laplace, quoted by Victor Guillemin, *The Story of Quantum Mechanics*, Charles Scribner's Sons, New York, 1968, p. 280.
41 James Gleick, *Chaos, Making a New Science*, Penguin Books, New York, 1988, p. 322.
42 Simon Singh, *The Code Book: The Science of Secrecy from Ancient Egypt to Quantum Cryptography*, Anchor Books, New York, 1999.
43 Piero E. Ariotti, *Benedetto Castelli's Discourse on the loadstone (1639–1640): the Origin of the Notion of Elementary Magnets Similarly Aligned*, Annals of Science vol. 38 (1981) 125–140, p. 135.
44 Piero E. Ariotti, *Benedetto Castelli's Discourse on the loadstone (1639–1640): the Origin of the Notion of Elementary Magnets Similarly Aligned*, Annals of Science vol. 38 (1981) 125–140, p. 137.
45 A. Wolf, *History of Science, Technology and Philosophy in the 18th. Century*, vol. 1, 2nd edition, George Allen & Unwin, London, 1962, p. 269.
46 Stephen T. Keith and Pierre Quédec, "Magnetism and Magnetic Materials", in *Out of the Crystal Maze, Chapters from the History of Solid-State Physics*, Eds. Lillian Hoddeson, Ernest Braun, Jürgen Teichmannn, Spencer Weart, Oxford University Press, New York, 1992, p. 364.
47 Stephen T. Keith and Pierre Quédec, "Magnetism and Magnetic Materials", in *Out of the Crystal Maze, Chapters from the History of Solid-State Physics*, Eds. Lillian Hoddeson, Ernest Braun, Jürgen Teichmannn, Spencer Weart, Oxford University Press, New York, 1992, p. 384.
48 F. J. Himpsel, J. E. Ortega, G. J. Mankey and R. F. Willis, *Magnetic Nanostructures*, Adv. Phys. 47 (1998) 511.
49 B. Willey, *The Seventeenth Century Background*, Penguin, Harmondsworth, 1967, p. 10–11.
50 P. W. Bridgman, "The Logic of Modern Physics", p. 37, quoted by B. d'Espagnat in *Reality and the Physicist*, Cambridge University Press, Cambridge, 1989, p. 89.

51 E. Nagel, *The Structure of Science*, Hackett Publishing Company, Indianapolis, 1979, p. 26.
52 W.C. Salmon, in "A Third Dogma of Empirism", in R. Butts and J. Hintikka, eds., *Basic Problems in Methodology and Linguistic*, 149–166, Dordrecht, D. Reidel, 1977, p. 166, quoted by W.C. Salmon, *Scientific Explanation and the Causal Structure of the World*, Princeton University Press, Princeton, 1984, p. 19.
53 C.G. Hempel in "Explanation in Science and History", in R.G. Colodny, ed., *Frontiers in Science and Philosophy*, 7–34, University of Pittsburgh Press, Pittsburgh, 1962, p. 10, quoted by W.C. Salmon, *Scientific Explanation and the Causal Structure of the World*, Princeton University Press, Princeton, 1984, p. 19.
54 G. Berkeley, "A Treatise Concerning the Principles of Human Knowlege [sic]", Dublin, Printed by A. Rhames, for J. Pepyat, 1710, §62, quoted by R. Torreti, *The Philosophy of Physics*, Cambridge University Press, Cambridge, 1999, p. 102.
55 R. Torreti, *The Philosophy of Physics*, Cambridge University Press, Cambridge, 1999, p. 103.
56 B. Willey, *The Seventeenth Century Background*, Penguin, Harmondsworth, 1967, p. 19.
57 P. Edwards in "Why", in *The Encyclopedia of Philosophy*, Ed. P. Edwards, Macmillan, New York, 1967, p. 297.
58 M. Heidegger, *An Introduction to Metaphysics*, New Haven, 1959, p. 4, quoted by P. Edwards in "Why", in *The Encyclopedia of Philosophy*, Ed. P. Edwards, Macmillan, New York, 1967, p. 297, p. 301.

Chapter 6

1 E.W. Cliver, Solar *Activity and Geomagnetic Storms: The First 40 Years*, Originally published in: *Eos, Transactions, American Geophysical Union, Vol. 75, No. 49*, December 6, 1994, Pages 569, 574–575.
2 R.T. Merrill, M.W. McElhinny and P.L. McFadden, *The Magnetic Field of the Earth*, Academic Press, San Diego, 1998, p. 7.
3 R.T. Merrill, M.W. McElhinny and P.L. McFadden, *The Magnetic Field of the Earth*, Academic Press, San Diego, 1998, p. 7.
4 Silvio A. Bedini, ed., "Compass", in *The Christopher Columbus Encyclopedia*, vol. 1, Simon & Schuster, New York, 1992, p. 207.
5 Luís de Albuquerque, "Contribuição das Navegações do séc. XVI para o Conhecimento do Magnetismo Terrestre", in *Estudos de História da Ciência Náutica*, Ministério do Planejamento e da Administração do Território, Lisboa, 1994, pp. 249–267.
6 David R. Barraclough, Geomagnetism: Historical Introduction, in *The Encyclopedia of Solid Earth Geophysics*, ed. David E. James, Van Nostrand Reinhold, New York, 1989, pp. 584–592.
7 R.T. Merrill, M.W. McElhinny and P.L. McFadden, *The Magnetic Field of the Earth*, Academic Press, San Diego, 1998, p. 7.
8 Alexander von Humboldt, quoted in Sydney Chapman and Julius Bartels, *Geomagnetism*, Oxford University Press, Oxford, 1951, p. 913.
9 J.L. Kirschvink, B.S. Jones and B.J. MacFadden, *Magnetite Biomineralization and Magnetoreception in Organisms*, Plenum Press, New York, 1985, p. 49.
10 A. Wolf, *History of Science, Technology and Philosophy in the 16th and 17th Century*, vol. 1, 2nd edition, George Allen & Unwin, London, 1962, p.302.

11 R.T. Merrill, M.W. McElhinny and P.L. McFadden, *The Magnetic Field of the Earth*, Academic Press, San Diego, 1998, p. 394.
12 D.P. Stern, A Millenium of Geomagnetism, Reviews of Geophysics, 40, number 3, 1–30 (2002), p. 13.
13 L.W. Widrow, Origin of Galactic and Extragalactic Magnetic Fields, arXiv astro-phi/0207240 v1 11 July 2002.
14 L.W. Widrow, Origin of Galactic and Extragalactic Magnetic Fields, arXiv astro-phi/0207240 v1 11 July 2002.
15 David L. Woods, *A History of Tactical Communication Techniques*, Martin-Marietta Corp., Orlando, 1965, p. 63.
16 R. Wiltschko and W. Wiltschko, *Magnetic Orientation in Animals*, Springer-Verlag, Berlin, 1995, p. 255.
17 F. Peterson and A.E. Kennely, *Some Physiological Experiments with Magnets at the Edison Laboratory*, NY Medical Journal 56 (1892) 729, quoted by John F. Schenk, *Safety of Strong, Static Magnetic Fields*, J. Mag. Res. Imaging 12 (2000) 2, p. 5.
18 R. Wiltschko and W. Wiltschko, *Magnetic Orientation in Animals*, Springer-Verlag, Berlin, 1995, p. 71.
19 R. Wiltschko and W. Wiltschko, *Magnetic Orientation in Animals*, Springer-Verlag, Berlin, 1995, p. 170.
20 Kenneth J. Lohmann and Sönke Johnsen, *The neurobiology of magnetoreception in vertebrate animals*, Trends in Neuroscience, vol. 23, no. 4, pp-1542–1548 (2000).
21 E.V. Mielczarek and S.B. McGrayne, *Iron, Nature's Universal Element*, Rutgers University Press, New Brunswick, 2000, p. 49.
22 James Phinney Baxter, *Scientists Against Time*, quoted by Ronald W. Clark, *Tizard*, MIT Press, 1965, p. 268.
23 C.J. Gorter, *Bad Luck in Attempts to Make Scientific Discoveries*, Physics Today, January 1967, pp. 76–81.
24 John S. Rigden, The Birth of the Magnetic-Resonance Method, in *Observation, Experiment, and Hypothesis in Modern Physical Science*, Eds. Peter Achinstein and Owen Hannaway, MIT Press, Cambridge, 1985, p. 231.
25 S.V. Vonsovskii, *Magnetism*, vol 1, John Wiley, New York, 1974, p. 10.
26 R.V. Pound, *From Radar to Nuclear Magnetic Resonance*, Reviews of Modern Physics vol 71 (1999) S54.
27 Manuel R. Mourino, *From Thales to Lauterbur, or From the Lodestone to MR Imaging: Magnetism and Medicine*, Radiology, 180 (1991) 593–612, p. 611.
28 Jerrold T. Bushberg, J. Anthony Seibert, Edwin M. Leidholdt, Jr. and John M. Boone, *The Essential Physics of Medical Imaging*, Williams & Wilkins, Baltimore, 1994, p. 312.
29 John F. Schenk, *Safety of Strong, Static Magnetic Fields*, Journal of Magnetic Resonance Imaging vol. 12 p. 2 (2000).
30 Marcus E. Raichle, *Visualizing the Mind*, Scientific American, vol. 270, April 1994, pp. 36–42, p. 41.
31 J.A. Jones, *NMR Quantum Computation*, Progress in Nuclear Magnetic Resonance Spectroscopy, vol. 38, 2001, pp. 325–360.
32 L.M.K. Vandersypen, M. Steffen, G. Breyta, C.S. Yannoni, M.H. Sherwood and I.L. Chuang, *Experimental Realization of Schor's Quantum Factoring Algorithm Using Nuclear Magnetic Resonance*, Nature, vol. 414, 2001, pp. 883–887.
33 Neil Gershenfeld and Isaac L. Chuang, *Quantum Computing with Molecules*, Scientific American, June 1998, pp. 50–55, p. 50.

Chapter 7

1 I. S. Jacobs, *Role of Magnetism in Technology*, Journal of Applied Physics vol. 40, 1969, pp. 917–928.
2 H. Hodges, *Technology in the Ancient World*, Penguin, Harmondsworth, 1971, p.17.
3 H. Hodges, *Technology in the Ancient World*, Penguin, Harmondsworth, 1971, p. 35.
4 André Leroi-Gourhan, "Le Complexe Technique du Néolithique", *Histoire Générale des Techniques*, Ed. Maurice Daumas, vol 1, pp. 61–74, Presses Universitaires de France, Paris, 1962, p. 69.
5 H. Hodges, *Technology in the Ancient World*, Penguin, Harmondsworth, 1971, p. 35.
6 R. F. Tylecote, *The Early History of Metallurgy in Europe*, Longmans, London, 1987, p. 92.
7 S. L. Sass, *The Substance of Civilization: Materials and Human History from the Stone Age to the Age of Silicon*, Arcade Publishing, New York, 1998, p. 83.
8 R. F. Tylecote, *A History of Metallurgy*, 2nd edition, The Institute of Materials, Brookfield, 1992, p. 10.
9 J. M. Roberts, *The Penguin History of the World*, Penguin Books, 1988, Harmondsworth, p.102.
10 R. F. Tylecote, *The Early History of Metallurgy in Europe*, Longmans, London, 1987, p. 52.
11 Georges Contenau, "Mésopotamie et Pays Voisins", *Histoire Générale des Techniques*, Ed. Maurice Daumas, vol 1, pp. 119–146, Presses Universitaires de France, Paris, 1962, p. 139.
12 J. M. Roberts, *The Penguin History of the World*, Penguin Books, 1988, Harmondsworth, p. 122.
13 Homer, *Odyssey*, ninth book, translation of Samuel Butler, The Great Books, vol. 4, Encyclopaedia Britannica, Chicago, 1978, p. 233.
14 A. C. Crombie, *The History of Science from Augustine to Galileo*, vol. 1, Dover Publications, New York, 1995, p. 140.
15 A. C. Crombie, *The History of Science from Augustine to Galileo*, vol. 1, Dover Publications, New York, 1995, p. 142.
16 A. C. Crombie, *The History of Science from Augustine to Galileo*, vol. 1, Dover Publications, New York, 1995, p. 222.
17 William Leiss, *Utopia and Technology: Reflections on the Conquest of Nature*, International Social Sciences Journal, vol. 22, no. 4 (1970), pp. 576–588.
18 Francis Bacon, Novum Organum, Part II, aph. 31, quoted by William Leiss, *Utopia and Technology: Reflections on the Conquest of Nature*, International Social Sciences Journal, vol. 22, no. 4 (1970), p.580.
19 Francis Bacon, in Alan Mackay, *Dictionary of Scientific Quotations*, Institute of Physics Publishing, Bristol, 1991, p. 20.
20 René Descartes, *Discours de la Méthode*, VI, translation in The Philosophical Works of Descartes, ed. E. Haldane and G. Ross, Dover, New York, 1955, vol. I, p. 119, quoted by William Leiss, *Utopia and Technology: Reflections on the Conquest of Nature*, International Social Sciences Journal, vol. 22, no. 4 (1970), p. 580.
21 Carolyn Merchant, The Death of Nature: Women, Ecology and the Scientific Revolution, Harper and Row, San Francisco, 1983, p. 169.

22 Karl Marx, *Capital*, volume 1, *Great Books*, vol. 50, Encyclopaedia Britannica, Chicago, 1978, p. 85.
23 John Stuart Mill, *The Oxford Companion to Philosophy*, Ed. Ted Honderich, Oxford University Press, Oxford, 1995, p. 608.
24 M. Blackman, *The Lodestone: a Survey of the History and the Physics*, Contemporary Physics 24 (1985) 319–331, p. 328.
25 Patricia Fara, *Sympathetic Attractions*, Princeton University Press, Princeton, 1996, p. 50.
26 K. J. Overshott, *Magnetism: it is Permanent*, IEE Proceedings A, vol. 138 (1), 1991, pp. 22–30, p. 24.
27 M. Blackman, *The Lodestone: a Survey of the History and the Physics*, Contemporary Physics 24 (1985) 319–331, p. 319.
28 M. Blackman, *The Lodestone: a Survey of the History and the Physics*, Contemporary Physics 24 (1985) 319–331, p. 328.
29 M. Blackman, *The Lodestone: a Survey of the History and the Physics*, Contemporary Physics 24 (1985) 319–331, p. 324.
30 K. J. Overshott, *Magnetism: it is Permanent*, IEE Proceedings A, vol. 138 (1), 1991, pp. 22–30, p. 25.
31 Patricia Fara, *Sympathetic Attractions*, Princeton University Press, Princeton, 1996, p. 79.
32 Correspondence of A. M. Tyndall to W. Sucksmith, 1941, quoted by Stephen T. Keith and Pierre Quédec, Magnetism and Magnetic Materials, in *Out of the Crystal Maze, Chapters from the History of Solid-State Physics*, Eds. Lillian Hoddeson, Ernest Braun, Jürgen Teichmannn, Spencer Weart, Oxford University Press, New York, 1992, p. 421.
33 Stephen T. Keith and Pierre Quédec, Magnetism and Magnetic Materials, in *Out of the Crystal Maze, Chapters from the History of Solid-State Physics*, Eds. Lillian Hoddeson, Ernest Braun, Jürgen Teichmannn, Spencer Weart, Oxford University Press, New York, 1992, p. 420.
34 J. D. Livingston, *The History of Permanent-Magnet Materials*, Journal of Metallurgy (JOM), February 1990, pp. 30–34, p.31.
35 U. Enz, Magnetism and Magnetic Materials: Historical Developments and Present Role in Industry and Technology, in *Ferromagnetic Materials*, vol. 3, Ed. E. P. Wohlfarth, Amsterdam, North Holland, 1982, pp. 1–36, p. 13.
36 U. Enz, Magnetism and Magnetic Materials: Historical Developments and Present Role in Industry and Technology, in *Ferromagnetic Materials*, vol. 3, Ed. E. P. Wohlfarth, Amsterdam, North Holland, 1982, pp. 1–36, p. 17.
37 J.M.D. Coey, *Whither Magnetic Materials?*, Journal of Magnetism and Magnetic Materials 196–197 (1999) 1–7, p. 3.
38 K. J. Overshott, *Magnetism: it is Permanent*, IEE Proceedings A, Vol. 138 (1) (1991) 22–30, p. 28.
39 J. D. Livingston, *The History of Permanent-Magnet Materials*, Journal of Metallurgy (JOM), February 1990, pp. 30–34, p. 33.
40 J. M. D. Coey, *Whither Magnetic Materials?*, Journal of Magnetism and Magnetic Materials 196–197 (1999) 1–7, p. 2.
41 Stephen T. Keith and Pierre Quédec, "Magnetism and Magnetic Materials", in *Out of the Crystal Maze, Chapters from the History of Solid-State Physics*, Eds. Lillian Hoddeson, Ernest Braun, Jürgen Teichmannn, Spencer Weart, Oxford University Press, New York, 1992, p. 416.

42 G. A. V. Sowter, *Soft Magnetic Materials for Audio Transformers: History, Production, and Applications*, Journal of the Audio Engineering Society, vol. 35, October 1987, pp. 760–777, p. 763.
43 G. A. V. Sowter, *Soft Magnetic Materials for Audio Transformers: History, Production, and Applications*, Journal of the Audio Engineering Society, vol. 35, October 1987, pp. 760–777, p. 764.
44 J. M. D. Coey, *Magnetism in Future*, Journal of Magnetism and Magnetic Materials, 226–230 (2001) 2107–2112, p. 2110.
45 S.A. Wolf, D.D. Awschalom, R.A. Buhrman, J.M. Daughton, S. von Molnár, M.L. Roukes, A.Y. Chtchelkanova, D.M. Treger, *Spintronics: a Spin-Based Electronics Vision for the Future*, Science, vol. 294, 2001, pp. 1488–1495.
46 J. D. Livingston, *The History of Permanent-Magnet Materials*, Journal of Metallurgy (JOM), February 1990, pp. 30–34, p. 33.
47 J. D. Livingston, *The History of Permanent-Magnet Materials*, Journal of Metallurgy (JOM), February 1990, pp. 30–34, p. 32.
48 K. J. Overshott, *Magnetism: it is Permanent*, IEE Proceedings A, Vol. 138 (1) (1991) 22–30, p. 28.
49 K. J. Overshott, *Magnetism: it is Permanent*, IEE Proceedings A, Vol. 138 (1) (1991) 22–30, p. 28.
50 I. R. Harris, Magnet Processing, *Rare-Earth Iron Permanent Magnets*, Ed. J.M.D. Coey, Oxford University Press, Oxford, 1996, pp. 336–380.
51 J. M. D. Coey, *Whither Magnetic Materials?*, Journal of Magnetism and Magnetic Materials 196–197 (1999) 1–7.
52 V. K. Pecharsky and K. A. Gschneidner Jr, *Magnetocaloric Effect and Magnetic Refrigeration*, Journal of Magnetism and Magnetic Materials, vol. 200 (1999), pp. 44–56.
53 C. Zimm, A. Jastrab, A, Sternberg, V. Pecharsky, K. Gschneidner, Jr., M. Osborne and I. Anderson, *Description and Performance of a Near-Room Temperature Magnetic Refrigerator*, Advances in Cryogenic Engineering, vol. 43, Ed. P. Kittel, Plenum Press, New York, 1998, pp. 1759–1766.
54 Marcel Proust, *Remembrance of Things Past*, vol. 1, Swan' s Way, translated by C. K. Scott Moncrieff and Terence Kilmartin, Vintage Books, New York, 1982, p. 51.
55 Philip Morrison and Phylis Morrison, *The Sum of Human Knowledge?*, Scientific American, vol. 279, July 1998, p. 95.
56 Philip Morrison and Phylis Morrison, *The Sum of Human Knowledge?*, Scientific American, vol. 279, July 1998, p. 95.
57 F. Jorgensen, *The Inventor Valdemar Poulsen*, Journal of Magnetism and Magnetic Materials, vol. 193 pp. 1–7, 1999.
58 K. W. H. Stevens, Magnetism, in *Twentieth Century Physics*, vol. II, eds. Laurie M. Brown, Abraham Pais and Brian Pippard, Institute of Physics Publishing, Bristol, 1995, p.1158.
59 J. D. Livingston, *100 Years of Magnetic Memories*, Scientific American, November 1998, p. 81.
60 M. H. Clark, *Making Magnetic Recording Commercial: 1920–1955*, Journal of Magnetism and Magnetic Materials vol. 193, 1999, pp. 8–10, p. 9.
61 J. D. Livingston, *100 Years of Magnetic Memories*, Scientific American, November 1998, p. 84.
62 M. H. Clark, *Making Magnetic Recording Commercial: 1920–1955*, Journal of Magnetism and Magnetic Materials vol. 193, 1999, pp. 8–10, p. 9.

63 M.H. Clark, *Making Magnetic Recording Commercial: 1920–1955*, Journal of Magnetism and Magnetic Materials vol. 193, 1999, pp. 8–10, p. 8.
64 M.H. Clark, *Making Magnetic Recording Commercial: 1920–1955*, Journal of Magnetism and Magnetic Materials vol. 193, 1999, pp. 8–10, p. 9.
65 Willem Andriessen, *'THE WINNER': Compact Cassette. A Commercial and Technical Look Back at the Greatest Success Story in the History of AUDIO up to Now*, Journal of Magnetism and Magnetic Materials vol. 193, 1999, pp. 11–16, p. 12.
66 J.D. Livingston, *100 Years of Magnetic Memories*, Scientific American, November 1998, p. 85.
67 C.J. Bashe, L.R. Johnson, J.H. Palmer and E.W. Pugh, *IBM's Early Computers*, MIT Press, Cambridge, 1989, p. 239.
68 Shan X. Wang and Alexander M. Taratorin, *Magnetic Information Storage Technology*, Academic Press, San Diego, 1999, p. 10.
69 K. O'Grady and H. Laidler, *The Limits to Magnetic Recording – Media Considerations*, Journal of Magnetism and Magnetic Materials, vol. 200, 1999, pp. 616–633, p. 618.
70 J.M.D. Coey, *Whither Magnetic Materials?*, Journal of Magnetism and Magnetic Materials 196–197 (1999) 1–7, p. 4.
71 Shan X. Wang and Alexander M. Taratorin, *Magnetic Information Storage Technology*, Academic Press, San Diego, 1999, p. 11.
72 IBM Research News, www.research.ibm.com/resources/news/20010518–whitepaper.shtml.
73 C.T. Rettner, S. Anders, T. Thomson, M. Albrecht, Y. Ikeda, M.E. Best, and B.D. Terris, *Magnetic Characterization and Recording Properties of Patterned $Co_{70}Cr_{18}Pt_{12}$ Perpendicular Media*, IEEE Transactions on Magnetics, vol. 38 pp. 1725–1730 (2002).
74 UC Berkeley's School of Information Management and Systems, www.sims.berkeley.edu/research/ projects/how-much-info-2003/execsum.htm.

Index

From Lodestone to Supermagnets. Alberto P. Guimarães

ISBN: 3-527-40557-7

b

c

d

h

i

j

k

l

m

q